# OUT OF THIN AIR

DINOSAURS, BIRDS, AND EARTH'S
ANCIENT ATMOSPHERE

# OUT OF THIN AIR

## PETER D. WARD

ILLLUSTRATIONS BY
DAVID W. EHLERT, MAMS, CMI

JOSEPH HENRY PRESS
WASHINGTON, D.C.

**Joseph Henry Press • 500 Fifth Street, NW • Washington, DC 20001**

The Joseph Henry Press, an imprint of the National Academies Press, was created with the goal of making books on science, technology, and health more widely available to professionals and the public. Joseph Henry was one of the founders of the National Academy of Sciences and a leader in early American science.

Any opinions, findings, conclusions, or recommendations expressed in this volume are those of the author and do not necessarily reflect the views of the National Academy of Sciences or its affiliated institutions.

### Library of Congress Cataloging-in-Publication Data

Ward, Peter Douglas, 1949-
 Out of thin air : dinosaurs, birds, and Earth's ancient atmosphere / Peter D. Ward.
   p. cm.
 Includes bibliographical references and index.
 ISBN 0-309-10061-5 (cloth)
 1. Dinosaurs—Extinction. 2. Dinosaurs—Evolution. 3. Paleoecology. 4. Paleobotany—Carboniferous. I. Title.
 QE861.6.E95W37 2006
 567.9—dc22

                    2006020396

Cover design by Van Nguyen.

Cover images: Jerry L. Ferrara and
        Francois Gohier / Photo Researchers, Inc

Printed in the United States of America.

*For Robert Berner, Navigator,*
 *who has so obviously relished the journey so far:*
 *You blazed a path to the old world*

# CONTENTS

## PREFACE

The highest peak of the great Himalayan mountains, Everest, tops out at 29,035 feet—nearly 6 miles above sea level. Its summit famously pokes into thin air, for at this height Earth's atmosphere has already thinned to pressures more akin to Mars's than to those found in the life-giving surface regions of Earth. Deadly, indeed, is that thin air—at least to those human climbers who dare enter it. But for other animals coming from the deep past, this is not the case.

At Everest's summit only a few humans can survive without oxygen augmentation and more often than not it was this factor that led to so many human deaths on this and other high mountains. There is sublime irony, then, in what may have been the last visions of some of those climbers as they lay dying. Surely some, staring upward into the pale blue regions where air pressure is even thinner than that which was killing them, watched the serene "V" of majestic geese flocks wending their way over the roof of the world, flying effortlessly thousands of feet above the Everest elevations that readily kill humans. Birds, it seems, care little for the physiological realities that rule us humans, us mammals, us creatures of ancestry far older than the birds themselves. Birds need far less oxygen than we ancient mammals. This book is, among other things, about why that is and how it came to be.

That birds can not only *exist* at altitudes lethal to mammals but in this thin, oxygen-poor air also perform the most extreme athletic exertion known in the animal kingdom—flying—is indeed a curiosity. But this is also a clue to a past that has only been recently discovered. The superiority of birds to any other group of vertebrates in terms of oxygen consumption has long been recognized. But it has been explained as a necessary adaptation to the athletic rigors of flying: birds need more oxygen because they exercise so hard and thus have evolved a more efficient lung system than ours. But this claim rings hollow, for the bird respiratory system found in the flyers is also found in non-flying birds. Something more fundamental seems to be at play. What if birds came from a time when even the surface of Earth, at sea level itself, had air as thin as that on Everest today? And if birds are so superior in thin air—what of their ancestors, the dinosaurs?

We think of this world—*our world*—as being how it has always been, even knowing that conditions on Earth in the far past must have been as if from some distant and alien planet. Yet so it was. And on alien worlds we expect alien organisms. Should we then be surprised if organisms from a different kind of Earth, an earlier Earth with a low-oxygen atmosphere, were so alien as to surely qualify for some science fiction epic? Ordinarily we have considered dinosaurs, indeed most extinct animals, as creatures made up of some assemblage of familiar traits. For the dinosaurs, these believed assembled traits are mammalian and avian characteristics that form animals looking different from anything alive today but acting in familiar ways. But we now believe that this was surely not the case, that dinosaurs and many invertebrates now gone were much different from the living creatures of today. Rather, many dinosaurs, cold-blooded yet active, were adapted for life in low-oxygen conditions through a superb new kind of respiratory system and moved with cadence and rhythm as alien as creatures in the movies. They were animals not familiar at all.

New discoveries in geology, geochemistry, and paleontology support this surprising claim and are the foundations of evidence that have led to the new history of life's pageant on this planet, which is the subject and reason for this book. Long ago, it seems, when thin air was the norm, when animals gasped and died of altitude sickness even at sea level, when no animals clung to even low mountain tops, a new brand of creature evolved with a superb new lung system. Breathing freely in the thin air, running down prey, or handily running away from lumbering predators that lacked this respiratory system, this stock of animals took over the world. We call them dinosaurs. Some of them, called birds, are famously still with us. This story also explains why the first common fossils in the Cambrian were segmented trilobites and why beautiful chambered cephalopods like our last holdout of that majestic race, the chambered nautilus, all but ruled the oceans for so long. And why creatures crawled from the seas, and why some then later crawled back in. All of these conclusions come from a new insight about the levels of that most necessary and most poisonous of gases, the giver and taker of life—oxygen.

## ACKNOWLEDGMENTS

It is not hard to know where to start in an otherwise usually difficult section to write: this book would not have been possible without the interest, ideas, and especially the criticism of one of America's greatest scientists, Dr. Robert Berner of Yale University. It is his seminal work that made this book possible, and if any good comes from the effort here, it is to Dr. Berner's credit. The errors that undoubtedly have crept in are mine alone, however.

I would like to thank the NASA Astrobiology Institute for funding the new science recounted in this book. I would like to thank the University of Washington and my colleagues in the Departments of Biology and Earth and Space Sciences for support during the writing of this book.

Many people helped with the text and with ideas. I am especially indebted to Tom Daniel, Ray Huey, Billie Swalla, and most of all Roger Buick. Ms. Nomi Odano kept affairs together in such a way that allowed me to write this book. I would like to thank my agent, Sam Fleischman, for help along the way, and the excellent editorial and production staff at Joseph Henry Press, especially the heroic efforts of Ms. Lara Andersen. David Ehlert did a magnificent job with art. Special thanks to Ken Williford for research and guidance.

# INTRODUCTION

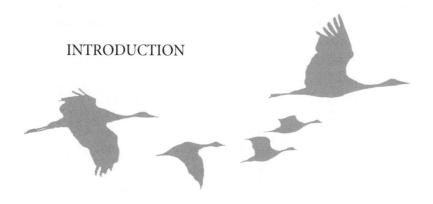

For practitioners of the vibrant new field of astrobiology, the study of life in the Cosmos, who peer ever outward into the universe, the Holy Grail is to discover at least one other Earth-like planet (and, of course, at least one other kind of life—alien life). Yet what exactly is an "Earth-like" planet? As anyone who has watched or read any science fiction knows, such a planet has water and a breathable-oxygenated atmosphere. But we are coming to realize that our view of an Earth-like planet usually means *one with an atmosphere similar to that found on Earth at the present-day.* Yet our current atmosphere is but a slice of a forever-changing entity and is greatly different from the atmosphere at most times in Earth's history. It is currently suited to us mammals—hence the high diversity of mammals alive today. But again, this has not always been the case. Two not-so-ancient versions, in astrobiological timescales, of our "Earth-like atmosphere" very nearly wiped out our furry ancestors some 250 million years ago and then tried again some 200 million years ago. If a small premammal named *Thrinaxodon*, whose delicate skulls have been collected in lowest Triassic strata, had not survived, what would life on Earth be like now? Perhaps we would have a diverse and unbelievably beautiful world of birds, in the air, on the ground, diving deeply into the sea, and perhaps they would be the dominant animals on Earth.

But we mammals did survive. The characteristics we now equate with "mammalness" are the consequence of these near-disasters during times of low-oxygen atmosphere, and also from times when oxygen was far higher than now. For most of our history we were the size of rats, sluggishly gasping in a low-oxygen world as we avoided the active dinosaurian overlords.

The thesis of the pages to come is deceptively simple: *The history of atmospheric (and hence oceanic) oxygen levels through time has been the most important factor in determining the nature of animal life on Earth —its morphology and basic body plans, physiology, evolutionary history, and diversity.* This hypothesis means that the level of oxygen influenced every large-scale evolutionary adaptation or innovation that is the history of animal life on Earth, that oxygen levels dictated evolutionary originations, extinctions, and the architecture of animal body plans. Support for this hypothesis will make up the chapters that follow.

## THE HISTORY OF LIFE

We begin a very brief version of the standard history of animal life on Earth with Charles Darwin. Darwin hoped that the history written in the fossil record would sooner or later support his contention about his then (1859) new theory of evolution: that change came about in small increments. In his *On the Origin of Species* he wrote:

> From the beginning of life on earth there was a connection so close and intimate that, if the entire record could be obtained, a perfect chain of life from the lowest organism to the highest would be obtained.

But this was hope, not history. A new understanding came a half-century later from Charles Wolcott, discoverer of the Burgess Shale, a Canadian rock deposit that gives our best look at the nature of what is known as the Cambrian Explosion, a short time more than 500 million years ago when most animal body plans rather suddenly appeared in the fossil record:

> In early times the Cephalopoda ruled, later on the Crustacea came to the fore, then probably fishes took the lead but were speedily out powered by the Saurians, the Land and Sea Reptiles then prevailed until Mammalia appeared upon the scene when it doubtless became a struggle for supremacy until Man was created.

Wolcott wrote this passage a century ago and since then much has been learned about the sequence of events making up the history of animal life. Recent books by Richard Dawkins (*The Ancestors Tale*) and Richard Fortey (*Life*) emphasize the history of life during the time of animals, from about 600 million years ago to the present-day. New discoveries about Earth's atmosphere, however, render even these most recent histories outdated.

The major events that all authors agree on as being consequential include (1) the origin of animals during the Cambrian Explosion, resulting in the first appearance of common animal fossils in the geological rock record of sedimentary strata; (2) the lower Paleozoic diversification in the seas, when a further expansion in numbers of marine species following the Cambrian Explosion resulted in a widespread shallow-oceanic shelf-fauna dominated by calcareous shelly organisms and the first coral reefs; (3) the mid- to late Paleozoic colonization of land, with successive and overlapping colonization by land plants, arthropods, vertebrates, and mollusks; (4) the late Permian mass extinction, an event that eliminated the majority of marine and land life and in so doing allowed a new mix of survivors to become dominant on land and in the sea; (5) the transitional world of the Triassic, a time on both land and sea when there were many new body plans appearing, including the first dinosaurs and true mammals; (6) the Jurassic and Cretaceous Age of Dinosaurs, including a characteristic marine fauna of shelled cephalopods and large flat clams in the sea; (7) the late Cretaceous diversification of modern-day land and sea life, followed by the enormous, asteroid-induced mass extinction that ended the Mesozoic Era; and (8) the Tertiary diversification of mammals. Each of these intervals or events in the history of life constitutes a chapter here.

Before beginning this examination, let's look at the concept of history and how it is studied and interpreted in the physical sciences—especially the history of life.

## EXAMINING HISTORY

There are four important questions in the study of any history: What? When? How? Why?

The first is the most basic: what happened? The *what* is the data of the story—in our case, the actual occurrences in the fossil or genetic record. The second, the *when,* is also straightforward, for it is the chronology that binds together the events of history into a linear, temporal succession. The third, the *how,* is often deeper and more difficult to learn than the previous two. It is the actual mechanism or motive that drove the historical record. Then, there is the lasting question—*why.* Sometimes it is very straightforward. More often that not, however, the underlying *why* of a history is the most difficult question to answer and is the subject of endless debate and revision for good reason. Along with being the most difficult question to answer, it is often the most interesting.

Stephen Jay Gould was clearly concerned as well with *why?* in studying the history of life, when in his book, *Wonderful Life,* he wrote:

> How should scientists operate when they must try to explain the results of history, those inordinately complex events that can occur but once in detailed glory? Many large domains of nature—cosmology, geology and evolution among them—must be studied with the tools of history. The appropriate methods focus on narrative, not experiment as usually conceived. The stereotype of the "scientific method" has no place for irreducible history.

But is the history of life—in this case, animal life—really irreducible? The scientific method, even laboratory experiment, is indeed not only relevant but also crucial to explain this particular history, the history of life. In various laboratories around the globe scientists are increasingly conducting experiments, many dealing with growing animals under varying oxygen content, to explain life's history.

So how do these questions apply to the history of animals and their great groupings of individual body plans that we call phyla? In terms of evidence, this history is a series of beginnings and endings: A particular group first appeared at a given time, flourished (or not), and finally died out (or continued into the present-day) in a linear, chronological order.

The *what* of this history is the identity of the organisms themselves and comes from two sources of data: from a zoology of the living that is based on two centuries of taxonomic study and more recently on decoding and comparing genetic sequences from groups of extant animals and from a paleontological study of the past over about the same two-century interval. Both of these endeavors have consumed the lives of men and women who have traveled to the ends of Earth to discover, classify, and catalog the extant and extinct biodiversity.

The *when* of this history has proven as difficult to discover as the *what* but for different reasons. Early geologists had no conception of the true age of Earth and had absolutely no way of discovering that great age. The one or two efforts to estimate the age of the planet, such as Lord Kelvin's innovative (and ultimately hugely wrong) effort to use the heat flow from Earth to estimate its age, yielded woefully low estimates. Discovering the age of individual fossils was even more impractical, and thus the chronology of life's history obtained from the fossil record was only a relative one. The succession of strata, one layer piled upon another, was slowly and painfully discerned by a century of stratigraphic study, eventually producing the geological timescale of eras, periods, and ages in use today. The first and last appearances of the major vertebrate groups that make up the *what* of life's history could, with the construction of the geological timescale, be put into relative succession. But the actual age of each event, such as the first appearance of mammals, or the last of dinosaurs, remained unknown until the inception of rock dating using radioactive age determination. Yet even without knowing the actual period in which mammals arose or dinosaurs died, paleontologists had, by the middle part of the 1950s, arrived at an accurate picture of animal evolution. While in the 50 or so years since then science has filled in innumerable details and developed a better understanding of the length of this history, in the larger view little has changed.

The third question of life's history—the *how*—has also remained little changed for a long time—since the nineteenth century, in fact. Charles Darwin, with his great theory on the evolution of organisms, gave an explanation as to why there has been a history of life. The diversification of the great vertebrate groupings came about through evolutionary processes. We obviously now know a great deal more about how evolution works than did the life historians in Darwin's day, but the overall explanation of how this works remains the same: evolution is responsible for providing the mechanism in this history of life.

We thus come to the final and most troubling question of history: the *why*? Why did the history of life unfold as it did? From the time of Darwin all the way until today there remains a group of nonscientists who find little trouble in answering this question: it has been God's will. But for those who choose to follow the methodology and philosophy of science, the *why* question concerning the history of life has been most troubling. Enormously interesting and complicated questions can be asked: Why does a group of organisms diversify? Why do we observe particular body plans and morphologies and what functions do they serve? Why are there so few (or so many) of any particular taxonomic unit, such as the orders or families of reptiles, or of mammals, for instance? Indeed, why are there mammals and reptiles at all? Why not some other biological reality? Why not truly flying fish, or water-breathing mammals, or even fire-breathing dragons, for that matter? Why are there no more dinosaurs, for instance? Now we can begin to see that the world of animals that now exists is just one of any number surely possible, at least theoretically. But for a particular battle in human history the map of nations would be radically different, and but for a particular disaster so too could the book of animals now on the planet be a radically different volume.

The animals of Earth are the way they are because of their history and because of the history of physical and biological conditions on Earth. They are as they are so as to be able to live on a planet with a given temperature range, a given overall pH, a given level of water, and given levels of other chemicals active in the biosphere. They are also as they are because of the history of events affecting the planet. New discoveries in our understanding of fundamental changes in the composition of Earth's atmosphere, especially its oxygen levels, through time require us to reexamine the accepted *whys* of animal evolution.

## AN OUTLINE OF CHAPTERS

This book is a reinterpretation of selected and important events and evolutionary breakthroughs during the past 550 million years, the time of animals, showing in chronological fashion why this author believes that it was the varied kinds of adaptation to varying oxygen levels that was the major stimulus to evolutionary change among the animal phyla. The evidence comes from both the history we have observed and what we know of animals living today. Following two introductory chapters, one discussing why animals need oxygen and the other recounting how various researchers have deduced the history of oxygen levels through time, the focus is on the new appearances and disappearances of various taxa and illustrating how specific adaptations among the various animal groups support the larger thesis of the importance of oxygen in forging evolutionary changes and results. Many of the evolutionary results are themselves newly hypothesized. By the end of this book, readers will come to appreciate the critical role of changes in atmospheric oxygen levels on the history of animal life on Earth and why there were dinosaurs and why there are birds.

*A late Triassic mammal-like reptile under predatory attack by two early dinosaurs. Here, the inferior lung of the mammal group is pitted against the superior lung system of the dinosaurs.*

# 1

## RESPIRATION AND THE BODY PLANS OF ANIMAL LIFE

I n the late eighteenth and early nineteenth centuries one preoccupation of naturalists (whom we would now call "scientists") was in classifying the hugely diverse assemblage of life on Earth into groups called "taxa." The basic unit of classification was called a "species," and it was defined based on members of the same species being able to interbreed. For many species however, the demonstration of successful interbreeding was impossible (especially for animals from the deep sea or from continents far from Europe, the site of most of this work and of course for all the fossil forms now extinct). For these latter animals classification into a species was based entirely on similarity of morphology. But it was recognized early on (especially by the eighteenth-century Swedish naturalist C. Linnaeus) that many different species, while not interbreeding, were so morphologically similar to others that all should be included in some "taxon" more inclusive than the species concept. Soon a hierarchical system was in place, with the following categories: species were members of a "genus," which in turn was a member of a "family," then "order," "class," "phylum," and, finally, "kingdom."

The distinction between each of these units was arbitrary unlike a species, which has an objective definition (interbreeding). However, one of the highest categories, the phylum, soon received its own defining aspect: members of a phylum shared a distinct suite of characters that could be called a "body plan." For example, all vertebrates have a backbone with a nerve cord, so this characteristic becomes the body plan of all vertebrates. All sponges show only two cell layers with a similar kind of cell allowing them to pump water through the body. This is another kind of body plan. Ultimately, the naturalists could find only 32 distinct kinds of body plans among animals and these became formalized later as the 37 animal phyla. Other examples of phyla are arthropods, all with a body plan with a jointed exoskeleton; *Cnidaria*, all with two cell layers and tentacles with stinging cells called nematocysts; and mollusks, with a soft body but a hard calcium carbonate shell at some time in their life.

The concept of the phylum has changed little since then. For example, one of the best modern descriptions of body plans comes from James Valentine in *On the Origin of Phyla*:

> The body plans of phyla have been much admired as representing exquisite products of evolution in which form and function are combined into architectures of great aesthetic appeal. While the phyla seem relatively simple in their basic designs, most contain branches that form important variations on their structural themes and some body types display remarkable embellishments in their morphological details. Presumably these variants reflect something of the ranges of ecological roles and environmental conditions in which the various phyla have evolved and functioned.

Why do the various phyla of life on Earth have the body plans they do? In other words, why did animals evolve the respective body plans seen today? Surely the answers lie in adaptations to conditions on Earth at the time of the evolution of the various animal body plans, no earlier than about 600 million years ago. It was a world different from today, with more oceans and less land surface, higher temperatures, more ultraviolet radiation, more atmospheric carbon dioxide, and less oxygen. There were no large predators or herbivores, and hence predation and competition to the emerging animals were only from their own kind. So what drove the body plans that emerged? *My contention is that respiration was perhaps the most important driver of animal body plans.* And yet in Rudy Raff's *The Shape of Life* and James Valentine's *On the Origin of Phyla*, two recent excellent treatments of this topic, respiration gets only a brief mention as a driving factor in the first and is not even listed in the index of the second. The emphasis here on respiration is due to new views about the nature of the atmosphere at the time of animal origins.

While many biologists have sought the origin of body plans comprising the various phyla, a question dealing with time, others have wondered why there are the number and kinds of phyla that there are. Why not 50 or 100 or just one? Why not animals with wheels? Stephen Jay Gould famously wondered if Earth would recapitulate its evolutionary history in approximately the same fashion—and thus end up with the same phyla—if we could somehow "replay the tape" of life on Earth. Given the same origin for our planet, would evolution play out through approximately the same history, or would forces of chance result in a wholly different assemblage of animals? While such an event could never take place, the question is more than academic. The new field of astrobiology is making clear that Earth is far from unique and that we can expect to find numerous Earth-like planets in space. How closely will Earth's history of life be replicated on any other world? In contrast to Gould's speculations, Cambridge paleontologist Simon Conway Morris has written that the process of convergence will re-create, at least under similar physical conditions, an assemblage of life forms that might appear wholly familiar to us. After all, in a medium of water and in an atmosphere with composition and pressure similar to that on Earth, the optimum ways to swim or fly—or respire—will, through naturally selected evolution, dictate the structures of organisms conducting these activities. But to what degree? Rudy Raff addresses this question when he states:

> We do not yet know the history of events that produced the body plans that appeared in the Cambrian. We don't know whether there is something inevitable about them.

So what shapes body plans? While it is clear that body plans have evolved for the multitude of functions required of any animal, is there any single function that is of overriding importance in design? It turns out that there might be. The fossil and genetic records strongly suggest that the various phyla first originated as bottom-dwelling marine organisms and that most, early on, evolved ways of moving across the substrate. Only later did animals evolve body plans that allowed burrowing within the bottom sediment and others that allowed swimming above the bottom. While there are many extant body plans that are not primarily involved in locomotion (sponges are a good example), most phyla are mobile either as adults or in early growth stages and, clearly, the need to move has affected the body design of many phyla. The integration of skeletons and locomotion also was important in resultant body plans.

Thus we have at least one answer as to why the phyla have the shapes and designs that they do: these were responses to the challenge of attaining movement, first on sea bottoms. We see clear manifestation of this in body plans incorporating streamlining and in the evolution of anterior and posterior regions on the body. But there are other body plans that initially seem to be related to movement but that actually are not. One of these is segmentation, where the body shows a repetitive grouping of smaller units, such as that seen in annelid worms and arthropods. Segmentation does allow a certain kind of movement such as in worms. And, with appendages attached, segmentation leads to other body plans. But is locomotion the primary reason for segmentation? Chapter 3 will return to segmentation and present a new hypothesis for its origin and use.

## OXYGEN, ENERGY, AND ANIMAL LIFE

Why do organisms bother with oxygen at all? Today there are huge areas of Earth, most underwater, that have little oxygen, so a body plan that helps an animal to live in low-oxygen environments would be very useful. But no animals use this kind of body plan. Why not? Aerobic respiration, the chemical reactions of metabolism in the presence of oxygen, yields up to 10 times more energy than does anaerobic respiration, a kind of respiration used by many bacteria. Fermentation is an example of how to get energy without oxygen. Complex life requires vast amounts of energy to meet its needs. For example, the formation of large molecules from smaller molecules in the synthesis of nucleic acids, lipids, and proteins involves a large input of free energy, just as the acquisition of energy requires the formation of specific cells or molecules, such as the chlorophyll molecule in plants—which takes energy. There is an old adage: it takes money to make money. This same idea can be analogized to energy.

While simple microbial life can derive its energy from a variety of sources, the choices decrease as life increases in size and complexity, simply because so much more energy is required. With the evolution of bodies capable of locomotion, the requirements increase yet again: an organism that moves can require 10 times more energy than a stationary organism. So a need of complex life (multicellular life, such as animal life) is a great deal of easily acquired energy. Only metabolisms using oxygen seem to give enough for animal life.

Where can such energy be found? Only in chemical bonds, it turns out. When some kinds of chemical bonds are broken, energy is released, and organisms have evolved ways to capture some (never all) of this released energy. There are many kinds of specific chemical bonds among the giant chemistry storehouse of naturally occurring compounds in the universe, but only a few have been harnessed by life. Of all combinations of elements in the periodic table, the hydrogen-fluoride and hydrogen-hydroxyl bonds yield the most energetic reactions per electron transfer, and this is because of the way hydrogen's electron is incorporated. (The specifics of this reaction are beyond the scope of this book: interested readers should refer to any text on physical chemistry). Of these, reactions with fluorine are marginally better than those with oxygen but fluorine has a huge disadvantage for life: it is biologically useless because it explodes when it comes into contact with organic molecules. The next best is oxygen; when present in an atmosphere or in water as dissolved oxygen it allows life to utilize the most energy-rich reactions commonly available.

There are two necessary functions of oxygen in aerobic life. The first, which accounts for most oxygen use, is to serve as a terminal electron acceptor in the energy production of aerobic metabolism—the energy extraction process that requires oxygen to "burn" sugar into energy. The second is its role in the biosynthesis of many enzymes necessary for life. More than 200 different enzymes are now known that require oxygen as part of their synthesis. Molecules that require oxygen include sterols, some fatty acids, and, most importantly, blood pigments necessary for respiration and the synthesis of mineral skeletons, such as shell or bone, which cannot be completed in the absence of oxygen for chemical reasons of bonding.

First, let's look at energy production. All cells on Earth need energy to run their cellular machinery. This energy comes from the chemical reaction that changes the molecule adenosine triphosphate (ATP) to adenosine diphosphate (ADP) and a free phosphorus atom through a pathway involving water (part of the water molecule is required to allow the chemical reaction to run). The splitting off of this single phosphorus atom releases energy that is utilized by the cell. There is no life on earth without ATP, and cells on Earth cannot live without a constant splitting of ATP into ADP. But organisms do not take in ATP. ATP has to be made, by bonding the phosphorus atom back on to ADP to make ATP. Organisms cannot either find or take ADP or ATP from other organisms by eating them. This involves an oxidation-reduction reaction. The ADP gains an electron and thus receives stored energy. At the same time, the electron donor is oxidized. While there is a wide range of chemical electron acceptors, the one that is most energetically favorable is oxygen and the species that use oxygen in this way. For animals the energy needed to start this ball rolling is the sugar called glucose. In the presence of oxygen, glucose is split, and the eventual product from several chemical reactions down the road is ATP. Organisms using this glucose plus oxygen chemical reaction to produce ATP from ADP (by splitting oxygen-hydrogen bonds) are said to use aerobic respiration. There is a chemical waste product of all of this—carbon dioxide.

Cells using this chemical pathway can make much more ATP than cells that do not over the same time period. For example, an animal that "burns" glucose by using oxygen makes more ATP per unit of time than does a bacterium using fermentation. All animals on Earth use this oxygen-mediated kind of cellular respiration. All animal cells thus need a constant supply of oxygen; without it they quickly die. Every body plan has its own way of getting the life-giving molecule, and this acquisition process is part of the foundation of any animal's design.

Now, let's look at other molecules of life that require oxygen, particularly those involved in oxygen and carbon dioxide movement to cells, called respiratory pigments. Respiratory pigments are used to help acquire oxygen from either air or water. They are formed from metal ions attached to organic compounds. *Hemoglobin* is the most familiar to us as it is the molecule/pigment that we use. It exists in a number of forms and is found throughout the animal kingdom in many (but not all) phyla, including vertebrates, echinoderms, flatworms, mollusks, insects, crustaceans, annelids, nematodes, and ciliates. Other pigments include *hemocyanin,* a copper-containing pigment found in gastropods, crustaceans, cephalopods, and chelicerates; and *hemerythrin,* the iron-containing pigment found in sipunculans, polychaetes, and priapulans.

All of these pigments bind oxygen more strongly when oxygen levels are high (in lungs or gills) and release it when oxygen levels are low (in respiring tissues). This occurs because in areas of low oxygen the chemical bond holding oxygen to the respiratory pigment molecule breaks more easily than it would in high-oxygen concentrations. Moreover, oxygen uptake is further facilitated in the respiratory structures where levels are low (and thus alkaline pH), and oxygen release is facilitated at actively respiring tissues where there is excess carbon dioxide and thus an acid pH.

We've seen that oxygen is critical, from a chemical perspective, in four necessary/useful functions of animal life. Indeed, we do not see animal life in oxygen-free zones.

## OXYGEN- AND ANIMAL-FREE ZONES

So how important is oxygen to animal and plant life on our planet? This question is perhaps best answered by looking at where animals do—and don't—live on Earth.

One of our most powerful methods in deducing the biology of any organism is in looking at the extremes that limit its life, for instance the upper and lower temperatures or chemical compositions an organism can withstand. In particular: at what level of oxygen are organisms never found?

It is not hard to find such places. On any beach, for instance, digging down into the sand with a shovel takes one quickly from the upper sandy layers, usually populated by a diversity of invertebrate animals, to a deeper, dark stratum that is both foul smelling and almost devoid of animal life. Usually it is necessary to dig no more than a foot or so to arrive at this darker regime, whose only animal life is made up of worms or clams with tubes or burrows that penetrate upward into the sea itself when the tide is high. If we continue to dig, we soon lose even these hardy colonists and find ourselves in a world devoid of animals. It is not devoid of life: rich colonies of microbial bacteria inhabit this world, but animals are nonexistent.

The blackness of this subterranean world tells us volumes about the chemical nature of this region. The black compounds and minerals found here are characteristic of a reducing environment as contrasted to an oxidizing one. A reducing environment is one in which oxygen is in short supply and any that somehow arrives is quickly involved in a chemical reaction. The reduction of chemical compounds in the sediment leads to the distinctive color, because many of the compounds contain carbon, which in substances like coal has a black color, and metals such as iron and lead, which in reduced states have minerals that are also dark in color, caused by the near absence of oxygen. It is the lack of oxygen that precludes animal life.

It is not just the sand that covers a reducing, anoxic world. Water itself can be virtually devoid of dissolved oxygen. Today there are large regions of the Gulf of Mexico that are composed of water volumes essentially lacking in oxygen, and the Black Sea in Asia, between Turkey and the old Soviet Union, is the largest known water body that has oxygen at very low levels. In both areas animal life is rare or absent, depending on the level of dissolved oxygen in the water.

We live in an oxygenated atmosphere. So how could places like the bottom of the Black Sea have little or no oxygen? It has an anoxic bottom because, unlike the larger oceans of the present-day, the Black Sea is composed of highly stratified, rather than mixed (from top to bottom), packages of seawater. Its relatively small size and more importantly, a low number of highly energetic storms or constant strong winds allows the seawater to settle into distinct strata based on their density. The lesson from the Black Sea is that, because of oceanographic conditions, even in a world with a highly oxygenated atmosphere there can be anoxic seas. And even large stretches of ocean considered "oxic" today could change. For example, because of phosphate- and nitrate-rich runoff from the Mississippi River valley region, large areas of the Gulf of Mexico are undergoing eutrification. This rather stagnant body of water is being fertilized to the point that large, oxygen-free water masses are becoming a yearly phenomenon.

Eutrification occurs when nutrients allow such lavish growths of plankton that all available oxygen in the water column is used up and cannot be replenished rapidly enough by ocean-air contact to avoid the formation of deadly, oxygen-free water masses. As these stagnant and anoxic body masses move into shallow waters, they kill off all benthic, or bottom-living, invertebrates, irrespective of whether they live on or in the sediment. Many invertebrates have very slow growth rates, and hence these relatively new oxygen-free zones are radically changing the nature of the sea bottom in important ways. Natural selection must be working in the Gulf of Mexico by favoring those organisms that have best adapted for living in low-oxygen conditions, but in the long history of life on Earth no animal has ever evolved a way to live in zero oxygen.

Low-oxygen water masses have been part of the oceans since the beginning of animal life. While low water-oxygen levels were exacerbated during times when the atmosphere itself had lower oxygen content than today, they are always present and thus there is always stimulus for selection for living in lowered-oxygen conditions. Today the oceans are mixed; the heat gradient between the warm tropics and cold polar regions creates the ocean circulation systems, which are composed of both surface currents and deeper, so-called thermohaline circulation systems, where cold, salt-rich, highly oxygenated bottom water is moved through the deeper oceans beneath warmer, fresher-water masses. This movement tends to oxygenate the oceans. But during past times of warmer climates, when there was much less of a heat gradient, the oceans were stratified, just as the Black Sea is today. There was a permanent presence of an oxygenated surface region atop an essentially worldwide, anoxic ocean at depth. Much of the Mesozoic was like this, and those of us who have collected Mesozoic marine rocks can attest to how widespread the dark anoxic ocean sediments from those times are. Usually such strata are fossil-free.

## OXYGEN IN WATER AND AIR

Because animals are obligated to harvest oxygen from the atmosphere or from a watery medium, they are affected by the varying concentrations of oxygen in air and water. There is no way that sufficient oxygen can be gained by eating food or drinking liquid with oxygen contained in it. And because animals are obligated to dispose of the waste products of respiration, chiefly carbon dioxide and water vapor, they are affected by the varying concentrations of carbon dioxide in air and water as well. What are the factors that influence the concentrations of oxygen and carbon dioxide in air and water?

First, water: At most there is only about 7 milliliters of oxygen per liter of water, but this is affected by many factors, including temperature and salinity (water pressure does not matter). The temperature of water plays a major role in the amount of oxygen that can be held dissolved in water. Colder water holds more oxygen than warm water. Because seawater contains such a large amount of dissolved solutes, there is less "room" among the water molecules for oxygen to squeeze in for gases to dissolve into seawater. At equal temperatures, then, a well-mixed body of fresh water will hold more oxygen than an equivalent volume of seawater. Well-mixed indicates that the water has the same characteristics from top to bottom, unlike, say, places such as the Black Sea as we saw above.

Oxygen and carbon dioxide also have different solubility in water. Carbon dioxide is far more soluble in water than is oxygen. However, even with this property there is never so much carbon dioxide in the water as to poison animals.

Now air: At present a liter of air contains 209 milliliters of oxygen at room temperature and at sea level. But with altitude, the amount of oxygen drops off. At the top of Mount Everest there is only one-third the air pressure that there is at sea level and hence the amount of oxygen is only a third. At any altitude the amount of oxygen diminishes as temperature rises.

To acquire an equal amount of oxygen, an animal living in liquid and getting its oxygen entirely from the liquid medium it lives in must process 30 times more seawater than an equivalent air breather. If breathing expends much energy for a particular animal, living in water is far more "expensive" than living in air, and at first glance it looks like life in air is a more promising physiology than life in water. But here the high dissolution rate of carbon dioxide in water comes to the aid of the water livers. Air breathers have a more difficult time ridding themselves of carbon dioxide from their blood than do their water-living cousins.

## OXYGEN AND ANIMAL RESPIRATION

The great differences in the behavior of oxygen and carbon dioxide in fresh and salty water compared to air have dictated the anatomy of many of the special respiratory structures found throughout the animal kingdom today and in the past as well. And another difference between the special respiratory structures used by water and air breathers is due to a fundamental law of chemistry. Both oxygen and carbon dioxide molecules are larger than a water molecule, so that any membrane that allows these gases to diffuse across it will also leak water. This has no consequence for an animal in water because if the animal becomes dehydrated, it can easily adsorb necessary water from its surrounding medium. However, in air the need to allow gas in lets water out. This leads to desiccation, a leading cause of death in both animals and plants to this day. Only in animals living in very moist environments, such as earthworms in moist soil, is this fundamental property inconsequential. For all other animals in dryer environments, some kind of impermeable body coating is needed to halt desiccation, but there have to be places where gas can come in as well.

Size also plays a part in the anatomy of respiratory structures. Very small animals can extract all the oxygen they need through passive diffusion, since if an animal is small enough its entire body is the respiratory structure. Five factors control any sort of diffusion: the solubility of the gas in question, the temperature at which the animal lives (or the body temperature it maintains), the surface area available for gas exchange, the difference in partial pressure of gases on either side of the respiratory structure, and the thickness of the barrier itself. Each of these factors has a say in dictating the anatomy of a respiratory structure, or, if the animal is small enough, the nature of body anatomy itself.

But because volume increases so much faster than surface area on an enlarging animal (or an evolutionary lineage undergoing a size increase), animals quickly move into size ranges that require special adaptation to acquire oxygen. The ratio of surface area to volume is a particularly insightful way to understand why special respiratory structures needed to evolve. The volume of a spherical organism grows according to diameter cubed, while its surface area grows according to diameter squared. Volume thus increases faster than surface area as diameter increases; the enlarging organisms soon reach the point where oxygen will not be able to enter all of the organism's cells fast enough.

Organisms can increase the critical surface area to volume ratios by changing their body plans from squat and compact to elongated. Aquatic forms with large surface areas include the numerous worm-shaped phyla. Similarly, changing the shape of the body to include numerous protrusions, fleshy spines, leafy additions, and even invaginations on the surface of the body increases the surface area to volume ratio. This class of adaptation is found in the gas exchange structures of both terrestrial and aquatic organisms. Most aquatic organisms, however, that rely on either basic body shape, small size, or simple evaginations or invaginations are limited to a life with limited locomotion, since, as we saw, locomotion requires great amounts of energy. They will be sluggish and slow moving and unable to support much in the way of nervous tissue, which, it turns out, is a large consumer of oxygen due to the many chemical reactions constantly taking place by signaling nerve cells. There is no better example than flatworms, small animals that are very sluggish and have only the most primitive kinds of nervous system. There is a difference in how much oxygen specific cell types require, and nervous tissue has the most voracious appetite for oxygen. Drowning victims die not from death of muscle or fat but from death of nerve cells. Thus, any group needing or evolving complex behaviors, such as locomotion or processing sensory input with their attendant nerve cells, very quickly has a need for an efficient source of delivering oxygen to nerve cells, among others.

## KINDS OF RESPIRATORY ORGANS

In general, respiratory organs are broadly classified as gills (evaginations), which are used in water, or lungs (invaginations), which are used in air. Respiratory organs evolved for water usually will not work in air and vice versa. The densities of the two media are so different that vastly different structures must be employed. That no animal has ever evolved an organ that could work both in air and water equally well suggests that it just cannot be done with biological material. We'll need to look at how gills and lungs acquire oxygen and dispose of carbon dioxide.

First, gills. As we have seen, if an aquatic animal is small enough relative to its oxygen requirements, no gill is needed. Oxygen simply diffuses across the outer body wall into the organism. But with greater size, this method is insufficient. Larger organisms are faced with some alternatives. The first is to produce tissue that has a high surface area to volume ratio and then connect this organ—this gill—to the rest of the body via some circulatory system. We can call this a *passive gill*, because it depends on contact with the medium to bring necessary oxy-

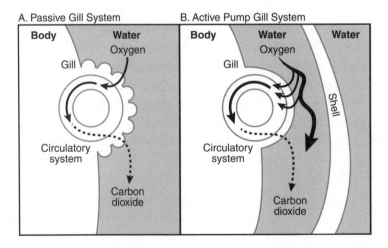

*Illustration of two kinds of gills. On the left is a passive gill: it is a located outside of the main body and depends on the passive uptake of oxygen in the surrounding water. On the right is a "pump gill", where the gill is enclosed by a shell, and water is actively pumped over the gill. This is a much better method for gaining oxygen from water.*

gen. It extends into the water, oxygen diffuses across its boundary into the blood or body fluid, and this oxygenated fluid is then passed inward to the interior cells. Salamander tadpoles, with their feathery gills, are a good example of this type of structure in vertebrates, while echinoids and nudibranch gastropods, among many others, are examples among the invertebrates. A disadvantage is that these types of gills are tasty targets for predators and, if nipped off, the wounded organism, assuming it survives the attack, would probably die anyway from lack of oxygen. Gills only work if there is a thin barrier between the water and the animal's blood and thus gills cannot be armored. Therefore, one common modification to this system is to place the gills within the body or build some kind of body armor over the delicate gills. Early mollusks used this trick, enclosing their gills (technically called ctenidia) with a space at the back of the body called a mantle cavity and then building a shell over the cavity and visceral mass. Crabs and many other modern arthropods also use this modification. Yet while hiding the gills out of predators' reach protects these delicate structures, it reduces the efficiency of respiratory exchange by reducing free exposure to water. Another problem is that the water that has already been processed for its oxygen could be reprocessed, thereby using up the organism's valuable energy and risking its oxygen starvation. To get around this, many animal groups evolved elaborate methods both to ensure that a sufficient volume of water is available to pass over the gill surface for the oxygen needs of the organism and to ensure that the water entering the gill region is not recirculated. For instance, while the shell of a gastropod, bivalve, or crab is typically explained as a compromise to allow defense and locomotion, in our discussion these same shells are considered part of the respiratory system, since some of their design specifically allows a more efficient or higher-volume passage of water over gills.

Two separate types of adaptations beyond simply increasing the surface area to volume ratio can increase the efficiency of gills. These are the second set of alternative approaches facing a larger aquatic animal. One increases the water flow over the gills; the other ensures that recirculation of already respired water is minimized.

Increasing the water flow available to the gills can be done in several ways. The simplest would be for the animal to position itself in high current areas, so that more water naturally passes over the respiratory structure. Many sessile invertebrates use this adaptation today, and certainly it must have been used in the past. A more direct adaptation is through *pump gills*, which actively use some mechanical means of pumping water across the gill membranes or respiratory exchange surface, such as by increasing the pressure differential between the water volumes on either side of the gill. At least three different sets of morphological adaptations would have to be evolved for this to work. First, there must be a space in the body in which to situate the gill so as to maximize the efficiency of water movement over the gill. Second, a circulation system must be evolved to maximize the gas transport and oxygen capture-efficiency of the gill itself—it is no use having a very efficient gill if the oxygen-laden blood cannot be carried to the cells of the body, often located far from the gill. Finally, some morphological structure must be built to maximize the flow of water across the gill through active pumping of some sort, such as through the evolution of cilia or flagella.

The variables that control oxygen uptake are the thickness of the tissue making up the gill surface, for this will control the adsorption rate of oxygen (more correctly the diffusion of oxygen across a biological membrane), and the size of the gill. The larger the surface area, the more oxygen that will be adsorbed. But since the amount of oxygen

Therefore, one way of categorizing gill types is to break out external from internal gill systems. An external gill, such as that in many salamander tadpoles, is in direct contact with water but does not have any mechanism for increasing the water flow over the gill, or for ensuring that already processed water is not recirculated over the gill. Those two capabilities, however, are possible in internal gills. If the gill is placed within some body or shell cavity, it is possible to create a pressure differential on either side of the gill that causes water to cross the gill surface at a higher rate than would occur if the gill is passively in contact with water. Such internal and "powered" gill systems are found in marine bivalves, which pump water into the space with the gills and then actively pump the deoxygenated, carbon dioxide–laden, respired water back out of the shell.

The variables that control oxygen uptake are the thickness of the tissue making up the gill surface, for this will control the adsorption rate of oxygen (more correctly the diffusion of oxygen across a biological membrane), and the size of the gill. The larger the surface area, the more oxygen that will be adsorbed. But since the amount of oxygen being acquired is a function of the rate at which water passes over the gill, the evolution of a water pump effectively increases the surface area.

Another gill adaptation to increase oxygen uptake is the use of "countercurrent" systems, used by organisms with very high metabolic rates, such as fish. Their high metabolic needs occur because of their exercise, which occurs during locomotion. Since most fish swim for some periods each day, they need more oxygen than similarly sized but unmoving animals. Countercurrent flows are the most efficient way to extract something from a fluid.

In fish gills, which utilize a countercurrent system, blood flows forward against the current of oxygenated seawater crossing the gill. In other words, as the fish brings in water across the gill, usually in a front-to-back direction, blood is pumped through the gill going the opposite direction—back to front. The blood being pumped across these gills has just come back from the body, and it is rich in bicarbonate ions that have been chemically changed from carbon dioxide to bicarbonate for the ride. When this bicarbonate-rich blood reaches the gills (or lungs) the bicarbonate is transformed back to carbon dioxide. There it encounters an environment with very little carbon dioxide. Because of this, it is pulled out of solution by diffusion and as a gas is expelled from the body.

As in gills, respiratory structures for air (lungs) involve the principle of exchanging oxygen and carbon dioxide. Often there is some morphological adaptation to force air at a higher pressure than atmospheric ambient pressure, and this entails the use of a "pump" of some kind. Animals do this in many ways. Our own solution is to use a series of muscles, the diaphragm, to inflate our lungs. Because air is far less dense than water, an equal volume of air contains many thousand times fewer molecules of oxygen than does the same volume of normally oxygenated water. Lungs therefore usually have much larger surface areas and, to do this, all are internal. Like powered gills, the animal uses some morphological mechanism to pump air into the lung chamber. Passive contact with air will not work for most larger animals and hence the many adaptations for "breathing"—the pumping of air into the lung chambers.

## RESPIRATORY SYSTEMS

A respiratory system can be defined as an assemblage of integrated cells and organ and tissue structures that deliver oxygen to the various cells in an animal's body and then remove carbon dioxide from the body. Respiratory systems are highly variable across the animal kingdom, incorporating specific morphologies involved in gas exchange, such as our own lungs and the gills of fish, but gas exchange is just the first part of the respiratory cycle. Oxygen has to be transported to the cells, and thus the animal's circulatory system is a major part of the respiration complex. Our blood, with its iron-bearing red blood cells, evolved specifically for oxygen and carbon dioxide transport, and the blood itself is classified as a tissue. The primary purpose of the heart in vertebrates, and larger invertebrates, is to take the blood or other medium, such as coelomic fluid, to every part of the body with speed. These fluids also take food material to the cells, so there is an overlapping of functions. And because the red blood cells are so important in respiration, we can say that their production, in the marrow of long bones in our body, is also a part of the respiratory system.

In the evolutionary pathway to an efficient respiratory system it is not just the primary oxygen acquisition organ that must be changed but also the entire system. Here is one example of just such a required change, one that we will revisit in more detail in Chapter 6 but that

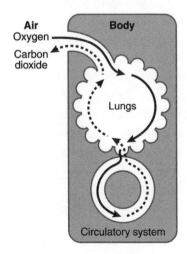

*Illustration of a lung system. Air goes into body, where oxygen and carbon dioxide exchange takes place.*

here can serve as an example of how complex it is to change a respiratory system. As we will see, the late Paleozoic was a time of rapidly falling oxygen, which diminished in volume relative to the gases in the atmosphere to the point that it seems to have affected the survivability of many groups of animals. This oxygen fall has been partially blamed for the great Permian extinction.

Therapsids, or mammal-like reptiles, were major victims of this extinction, but one genus of this group not only survived the extinction but actually flourished. The genus *Lystrosaurus*, which appears in the final throes of the mass extinction, differed from other therapsids in having an enormous, barrel-like chest with larger lungs. The hypothesis is that this group evolved larger lungs in response to dropping oxygen levels and thereby survived the oxygen fall. But increasing lung size would not have been the only evolutionary adaptation. The larger lungs, with their new size and shape, would necessitate changes in the morphology of the blood vessels servicing the lungs. Even if these changes involved only repositioning of the aorta and major blood vessels going to the lungs, there would necessarily have been wholesale changes in chest morphology, probably involving the heart as well. And there may have been even more radical and extensive adaptation in response to this, perhaps including the first evolution of endothermy, or warm-bloodedness, as evidenced by changes in the series of bones in the nose that allow warm air to go in and out of the body and that at the same time reduce water loss caused by breathing. In the entire history of the therapsids, themselves consisting of hundreds of individual species, the evolutionary changes resulting in the origination of *Lystrosaurus* in the late Permian represent the most radical morphological change in the formation of a new genus known from the entire history of the group. Evolution happens all the time in animals, and many of the changes commonly seen are overall size increase or, among mammals, changes in tooth morphology. But here it is suggested that respiratory changes usually require the most extensive morphological transformations.

In the example of the Permian-Triassic *Lystrosaurus*, increasing the oxygen-acquisition process can involve increasing the size of the lungs. But other respiratory system adaptations are to increase breathing rate or to modify the fashion by which air is delivered to the lungs. As we will see, both adaptations have appeared in land animals in times of low oxygen.

What specific kinds of respiratory systems are observable in animals, and how have they changed over time? How can these various kinds of respiratory systems be compared one to another? The hypothesis that changing oxygen levels through time provoked the most radical evolutionary changes in animals of all evolutionary stimuli would be supported if it could be shown that those animals with more efficient lungs were more successful (at the level of individual survival anyway; "successful" has many meanings and could equally refer to success as measured by diversity as well) in times of low or lowering oxygen than those with less efficient respiratory systems.

The problem of relating efficiency of respiration to successful survival is a difficult one, made even more vexing by the fact that most animals double up functions. For instance, bivalved mollusks use their gills for both respiration and food acquisition. If we see changes in the gills over time, are these respiratory adaptations or increases in feeding efficiency? We will return to this question, but first let's note a simpler case. Our own lungs provide such a case: they are used exclusively for oxygen acquisition and carbon dioxide excretion. This is typical of vertebrates, but, as we shall see, respiratory structures often double as feeding organs in many invertebrates. First, we need to quantify respiratory system efficiency.

One way of measuring efficiency of a respiratory system is to compare the rate at which oxygen is captured from either water or air and then transported to the body. There is a good deal of literature in the field of physiology that has been concerned with this question, so that much is known at least for many groups of vertebrates. Unfortunately, far less is known about invertebrates.

One well-known result involves the relative efficiency of mammals and birds. At rest the bird respiratory system, as measured by the amount of oxygen delivered to the body over a given amount of time, is at least 33 percent more efficient than any mammal lung. The lungs of high-altitude South American mammals such as Vicuna and Alpaca are among the best that mammals have mustered, but even they pale when compared to the efficiency of even low-altitude bird lungs. (It should be noted that this kind of study of birds is still in its infancy, with many questions on the relative efficiency of lungs found in various-sized birds and from different habitat types just beginning.)

How can this difference be explained? While perhaps the blood of mammals and birds is sufficiently different to allow the avians to transport oxygen more efficiency, the likely cause is the radically different lung morphology that birds have compared to mammals.

Fundamentally, it is the amount of dissolved oxygen in the blood that is to be compared. Physiologists insert instruments into the blood (ow!) and directly measure how much oxygen is present. In many respiratory organs this value is related to the rate and volume of oxygen-bearing medium that comes in contact with the blood at the gas absorption sites. It is this process that can be most easily shaped and changed by evolution.

To highlight the various possibilities among existing and some extinct organisms, a table of respiratory organs for various body plans is shown in the Appendix at the back of the book. The appendix is but a short list of taxa, mostly generalized at higher taxonomic levels, but it illustrates the great variety of respiratory styles across the animal kingdom. As we profile the various groups of animals making up the specific events in the history of life, we can ask whether a specific group first evolved in a high- or low-oxygen world (compared to the present-day) and if respiratory adaptations of any kind can be observed during or immediately after times of oxygen content change in the atmosphere, either up or down. The difficulty is that respiratory organs are composed of mainly nonpreservable soft parts. However, because of the high degree of integration of skeletal parts and respiration found in so many groups, respiratory adaptations often can be inferred or directly observed.

## MOLECULE OF ANIMAL LIFE

Oxygen is thus essential for animal life, and animal life consequently has evolved a host of structures for its acquisition. Because Earth has widely varying oxygen concentrations in different environments, some of the differences shown in the Appendix are related to living in different oxygen regimes. But we must take into account a new variable as well. Not only does oxygen concentration vary from place to place, it has also varied through time. It seems possible that an organism's type of respiratory system could be related to the oxygen levels present on Earth when that organism first evolved. Some animals first appeared when atmospheric oxygen was much higher than now, some under just the opposite condition. Chapter 2 looks at the history of oxygen in Earth's atmosphere and oceans and how this molecule's history has been studied.

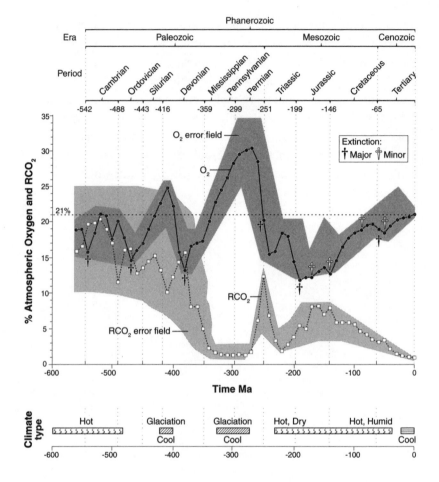

## 2

## OXYGEN THROUGH TIME

W                e live today in a world that is quite atypical of Earth over most of its 4.6 billion years long history—and probably atypical as well of much of Earth's future history. Yet it is human nature to regard the world that we are used to—what might be called "the world as we know it"—as a permanent thing. The oceans, prairies, and mountain chains—even the air we breathe—seem the norm and therefore permanent things. This sense of permanence is especially true for our atmosphere, since it has a multitude of functions that are necessary to allow the continued existence of life: it distributes water, drives critical chemical cycles, feeds plants, provides oxygen for animals, shields the surface from deadly ultraviolet light, and moderates global temperatures. Surely, it would seem, its levels should have remained constant at least from the time that animals first appeared on our planet, some 600 million years ago. But not so far back in Earth's history, perhaps only 5 million years ago, oxygen levels were significantly higher than now, while less than 100 million years ago, oxygen levels were significantly lower than today. The atmosphere has changed markedly over time, and these changes may be key to understanding major evolutionary changes of life on Earth over time, including changes of the vertebrates. This chapter will look at how our current atmosphere came about and how it was recently discovered that the air has undergone relatively recent (compared to the age of

Earth) periods of hypoxia or low-oxygen, whereas the bottoms of the ocean have undergone not only hypoxia but, on occasion, anoxia, or the complete absence of oxygen, and also periods of higher than present oxygen.

## THE ATMOSPHERE AND THE GLOBAL OCEAN

There can be no discussion about the atmosphere without including the global ocean. The two are coupled; even small changes in the temperature or chemistry of the global ocean can produce enormous changes in the atmosphere.

The composition of Earth's present-day atmosphere is basically known. Essentially it is made up of two gases: 78 percent of its volume is nitrogen, and 21 percent is oxygen. The remaining 1 percent is made up of trace amounts of other gases. Yet even at this small volume, this 1 percent has a huge effect on the planet, for within this 1 percent are both the important greenhouse gas of carbon dioxide and water vapor (itself a gas). "Greenhouse gas" is the term now used to describe any gas at work to trap heat in the atmosphere, thus warming the planet. How long has our planet had this atmosphere?

The atmosphere of our planet is as old as Earth itself. The two originated at the same time somewhere around 4.6 billion years ago— a date that is almost one-third the age of the Universe itself. The planet was molten soon after formation, but rapid cooling set in and as temperatures dropped, the planet rapidly evolved. Once formed, the solid Earth and its gaseous atmosphere evolved in quite different ways, even though each influenced the other over time. Like all planets, Earth formed through accretion of particles in a solar or planetary nebula. The formation of Earth was but one part of the formation of the entire solar system. As our planet accreted, it began to differentiate, with the heavier elements sinking toward the center and the lighter elements staying near the surface. In this fashion the major structural elements of our planet, its dense inner core, middle mantle, and outer crust regions formed. This process led to rapid changes in the atmosphere of the forming earth as well. Enormous quantities of gas were trapped in the differentiating Earth and sequestered far beneath the surface of the planet. Over time this gas began to escape into the atmosphere and in so doing rapidly changed the composition of the planet's gaseous envelope. We have a clue about the nature of the gas still trapped within Earth by studying the gas composition of volcanoes. Present-day composition of volcano effluents are 50-60 percent water vapor, 24 percent carbon dioxide, 13 percent sulfur, and about 6 percent nitrogen, with traces of other gases, a composition that differs markedly from the current atmospheric composition.

Our world ocean (since it is interconnected, even though we give parts of it separate names) has also changed its chemistry, mainly by changes in salinity through time. Most scientists believe the oceans have gradually become saltier through time, although a smaller but vocal group advocates that the oceans have become less salty through time. (The amount of salt in the oceans has no effect on the atmosphere and thus plays no part in our story.) The most characteristic aspect of our planet is its envelope of liquid water, and it would seem reasonable to assume that the voluminous oceans of planet Earth were created as part of the natural evolution of the cooling planet. This may not be the case, however. While the outer planets and moons of our solar system, from Jupiter outward, are rich in water, astronomers modeling how solar systems form have discovered that water should be in short supply among the inner parts of the solar system. Because of this, it is now believed that an appreciable volume of Earth's surface water was brought here from the outer reaches of the solar system by comets impacting the planet early in its history. If this is the case, it indicates that much of our oceans and perhaps an appreciable portion of our atmosphere are exotic to Earth. Most of this delivery happened in the first 500 million years of Earth's history, and the rain of comets onto the planet during the period from 4.2 billion to 3.8 billion years ago, known as the Heavy Bombardment period, may have caused Earth's early oceans to be repeatedly vaporized into steam.

The composition of Earth's atmosphere early in its history is a controversial and heavily researched topic. While the amount of nitrogen may have been similar to that of today, there are abundant and diverse lines of evidence indicating that there was little or no oxygen available. Carbon dioxide, however, would have been present in much higher volumes than today and this carbon dioxide–rich atmosphere would have created hothouse-like conditions through a super greenhouse effect, with carbon dioxide partial pressures (measured as the actual amount of total gas pressure exerted by the atmosphere) 10,000 times higher than today.

There is abundant evidence that the present-day atmosphere is very different from that of the past. The most compelling lines are geologic. Today, the atmosphere contains so much oxygen that reduced metal species quickly oxidize: the familiar rusting of iron to a red color or the oxidation of copper to shades of green is evidence of this. In similar fashion, many metal-rich or organic-rich types of sediment quickly bind with atmospheric oxygen to produce oxidized minerals. Long ago in Earth's history, however, minerals formed that are no longer seen on the planet's surface. Before about 2.5 billion years ago the formation of "red beds," sedimentary beds rich in oxidized iron minerals such as hematite did not form. Instead, there was formation of "banded iron formations," composed of only partly oxidized iron species. Other rock types from this ancient time include uranium oxides and iron pyrites that cannot form in today's atmosphere. This evidence strongly suggests that prior to 2.2 billion years ago there was no free oxygen in the atmosphere and little oxygen dissolved in seawater.

Even though there must have been, at most, only a few percent of oxygen in the gases making up Earth's atmosphere as late as 2.2 billion years ago, soon after that the amount of oxygen began to climb rapidly. Where did all the oxygen come from? Some oxygen can be generated by photochemical reactions, where water high in the atmosphere is broken by sunlight into hydrogen and oxygen, but this process could account for only a small percentage of the oxygen rise. The most likely explanation is that most came from photosynthesis by single-celled bacteria. Life is known to have evolved on Earth by about 3.5 billion years ago, perhaps hundreds of millions of years earlier than that. Certainly, by 3.5 billion years ago, life had evolved to the point where cyanobacteria (informally and improperly known as blue-green algae) were widespread in the oceans.

The cyanobacteria were the first organisms to use carbon dioxide to produce free oxygen. They still exist and use carbon dioxide as a source of carbon for building cells. They cannot use it for energy. They also made nitrogen available to their protoplasm by developing specialized structures (*Heterocysts*) as locations for nitrogen fixation. The cyanobacteria were eventually co-opted by other, larger cells (the eukaryotic cells that contained a distinct, membrane-bounded nucleus, in contrast to the smaller bacteria without a nucleus). This theory, known as the Endosymbiosis Theory, was proposed by biologist Lynn Margulis. Some members of the cyanobacteria became the modern chloroplast, the part of the plant cell in which photosynthesis is carried out. This transition to larger "plant" cells took place perhaps 2.7 billion years ago, and by 2.3 billion years ago a buildup of oxygen in the atmosphere was taking place.

The buildup of oxygen in Earth's atmosphere led to the formation of an ozone layer thick enough to shield life on the surface of the planet from the harmful effects of ultraviolet radiation. Ozone is another chemical form of oxygen. Because of its different bonds, it cannot be used to "burn" sugars but does screen out harmful radiation that would otherwise hurt organisms on Earth. The amount of oxygen depends in part on the amount of oxygen in the atmosphere. At times of low-oxygen, all the oceans will similarly have little oxygen in them. However, since the amount of oxygen that can dissolve into seawater is also affected by temperature, as shown in the previous chapter, warm oceans might have little oxygen in them everywhere but in a narrow surface zone, despite there being high-oxygen levels in the atmosphere. For instance, this condition exists in the modern Black Sea.

### ATMOSPHERIC OXYGEN ESTIMATES OVER TIME

The amount of oxygen in Earth's atmosphere is determined by a wide range of physical and biological processes, and it comes as a surprise to most people that the level of oxygen in the atmosphere fluctuated significantly until relatively recently in geological time. The same is true of carbon dioxide. As we read in the press virtually daily, carbon dioxide levels can be raised quickly if sources of the gas start putting more of it into the atmosphere by, say, volcanoes or SUVs. But why do the levels of these two gases change on a slowly aging planet such as Earth?

There is no simple answer to questions about the cause of either rises or falls in the amount of oxygen in the atmosphere, much as we would like there to be. For more technical details on the various geochemical and geological issues underlying the discussion that follows, Robert Berner's book, *The Phanerozoic Carbon Cycle*, is highly recommended.

The major determinants of the changes in atmospheric oxygen levels are a series of chemical reactions involving many of the elements abundant on and in Earth's crust, including carbon, sulfur, and iron. The chemical reactions involve both oxidation and reduction, processes that involve chemical reactions where certain elements either add or lose electrons. In the case of oxidation reactions, free oxygen (oxygen) combines with molecules containing carbon, sulfur, or iron to form new chemical compounds and in so doing oxygen is removed from the atmosphere and stored in the newly formed compounds. Oxygen is liberated back into the atmosphere by other reactions involving reduction of compounds. This is what happens during photosynthesis as plants liberate free oxygen as a by-product of the break-up of carbon dioxide through a complex series of intermediate reactions. Two important cycles ultimately dictate oxygen levels: the carbon cycle and the sulfur cycle. There may be other elements that are important as well, but currently they are deemed far less instrumental in affecting oxygen levels than are carbon and sulfur.

Let's look first at the sulfur cycle. Sulfur is found in a wide variety of compounds, but the most important for understanding the rise and fall of oxygen over time is pyrite. This gold-colored cubic mineral is familiar to us as "fool's gold," and while of little value compared to gold in monetary terms, it is hugely important in dictating the amount of oxygen in the atmosphere and hence the state of the biosphere. Sulfur is added to the oceans from rivers as it weathers out of pyrite-bearing rocks on the continents, or it comes from sulfur-rich sedimentary rocks, such as gypsum and anhydrate. These latter are already in chemical states that do not react with oxygen. Such is not the case with pyrite, however. There are huge quantities of pyrite locked in a variety of rocks, most importantly dark shale that originated in the oceans, which are uplifted onto continents via plate tectonic mechanisms and then weathered under the onslaught of rain, wind, cold, and heat. There are also great piles of the two sedimentary rocks, gypsum and anhydrate, that in similar fashion weather and release oxidized sulfur compounds into rivers and ultimately the sea.

There is a second pathway as well for the oxidation of sulfur compounds, one that takes place deeper in Earth, as pyrite is broken down during burial or subduction of pyrite-bearing rocks in the descending slabs of rock at the deep-sea trenches. Eventually this pyrite is heated to the point that it combines with oxygen, and the products of the reaction are emitted as the familiar and noxious-smelling sulfur gases found in volcanoes and hot springs, the poisonous gases called hydrogen sulfide and sulfur dioxide. When this happens, oxygen levels can drop in the atmosphere, especially if Earth is undergoing a phase of mountain building that exposes vast new reserves of sulfur-bearing rocks to erosion.

Even more important in dictating oxygen levels is the carbon cycle. Carbon makes up much of our bodies. Whether large quantities of reduced carbon compounds, such as animal and plant bodies after death, are left on the surface of the planet to react with atmospheric oxygen or are quickly buried has a major effect on oxygen levels. The rate of burial of organic carbon, along with the burial rate of sulfur-bearing compounds, is thus the major determinant of atmospheric oxygen levels.

Unfortunately, there is no direct way to measure past oxygen levels. (About a decade ago one such method was thought to be discovered: trapped air in amber was ballyhooed as a direct measure of past oxygen levels—until it was found that the small bubbles were not cut off from later changes in atmospheric levels.) Indirect methods based on an understanding of the relative ability of various minerals to undergo chemical changes in the presence or absence of oxygen have been used to infer relative oxygen levels, as have indirect methods based on biological evidence. For instance, in South Africa a very old mineral deposit was found that contains sedimentary uranium minerals. These minerals quickly change into other kinds of minerals when exposed to oxygen. But their presence in river deposits of more than 3 billion years ago yields powerful evidence that the land of the time (where rivers are) was covered with an oxygen-free atmosphere.

Another method involves computer modeling of past oxygen and carbon dioxide levels through time, based on a set of equations and then checking these model values with the mineralogical or paleontological evidence to validate the models. There have been a number of models specifically derived to deduce past oxygen and carbon dioxide levels through time, with the set of equations for calculating levels of carbon dioxide referred to as "GEOCARB" being the most elaborate and oldest. This model and a separate model for calculating oxygen have been developed by Robert Berner and his students at Yale University.

## GEOCARBSULF

The GEOCARBSULF model is a recent combination of the much earlier models for carbon dioxide (GEOCARB) and oxygen (isotope mass balance model). It is a computer model that takes account of the many factors thought to influence atmospheric oxygen and carbon dioxide.

A computer model such as GEOCARBSULF must take account of "forcings," processes that affect the oxygen levels. Chief among these are the rate of metamorphic and volcanic degassing of reduced carbon- and sulfur-containing gases, the rate of mountain uplift, sea level change, burial of organic matter accompanying land plant evolution, and colonization of land by plants. Each of these factors influences the burial rates of reduced carbon and pyrite sulfur (which cause atmospheric oxygen to increase) and the rates of erosion and thermal decomposition to volcanic gases of sedimentary rocks containing significant quantities of reduced carbon and pyrite sulfur (which cause oxygen to decrease).

Understanding the history of oxygen through time thus involves understanding the causes for the rise and fall of oxygen, and thus it is imperative to understand when and if burial and weathering rates of organic carbon and pyrite were either enhanced or inhibited. Carbon dioxide levels are also influenced by these factors. They are also influenced by the enhancement of weathering by land plants; the apportionment of carbonate burial between deep and shallow seas; and the change of insolation (the amount of sunlight hitting the planet through time), which is as much as a 4.4 percent increase of solar energy hitting Earth from the earliest Cambrian to the present-day. The end result of all of this is an estimate of oxygen level. If this is done for various times in Earth's history, a graph is produced. Results of the computer model for oxygen as a percentage of the atmosphere over time are shown in the figure that begins this chapter.

The so-called Berner curves of oxygen levels through time come from the iterations of the GEOCARBSULF and earlier oxygen models. They are not the only models, though. Recently, geochemist Noah Bergman and his colleagues Timothy Lenton and Andrew Watson published a different estimate for oxygen and carbon dioxide through time and at the same time included calculations of global temperatures through time. As with GEOCARBSULF, their model, called COPSE, depends on the input of values for many variables that are known to or are suspected of controlling oxygen and carbon dioxide levels in the atmosphere. Their results, while showing somewhat different levels of oxygen and carbon dioxide through time compared to Berner's results, do, in fact, show the same shape of curve and thus seem to corroborate the Berner results.

## GEOLOGICAL TIME

If oxygen has changed through time, when was it high and when low? To discuss the model results for atmospheric oxygen levels over time, we thus need to refer to the geological timescale. Because this book is about history, at this point we need to briefly digress from our story and look at how geological time has come to be known.

The development of a geological timescale was the product of two centuries of investigation. The geological timescale began as a table of strata, with older strata beneath younger. Gradually the fossil content of the strata defined the units. Only in the past half-century, when radiometric age dating methods were discovered and applied, did the geological timescale acquire ages in absolute time units of years.

One of the fundamental divisions of the timescale is made by the presence or absence of common animal fossils. We now know that animals first appeared about 540 million years ago. The entire sweep of

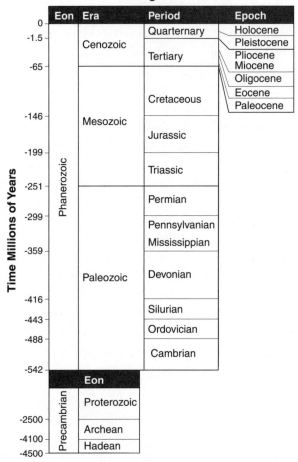

**Geologic Time Scale**

| | Eon | Era | Period | Epoch |
|---|---|---|---|---|
| 0 | | | Quarternary | Holocene |
| -1.5 | | Cenozoic | | Pleistocene |
| | | | Tertiary | Pliocene |
| -65 | | | | Miocene |
| | | | | Oligocene |
| | | | Cretaceous | Eocene |
| | | | | Paleocene |
| -146 | | Mesozoic | | |
| | | | Jurassic | |
| -199 | Phanerozoic | | | |
| | | | Triassic | |
| -251 | | | | |
| | | | Permian | |
| -299 | | | Pennsylvanian | |
| | | | Mississippian | |
| -359 | | | | |
| | | Paleozoic | Devonian | |
| -416 | | | | |
| | | | Silurian | |
| -443 | | | Ordovician | |
| -488 | | | | |
| | | | Cambrian | |
| -542 | | | | |

| | **Eon** | | |
|---|---|---|---|
| | Precambrian | Proterozoic | |
| -2500 | | | |
| -4100 | | Archean | |
| -4500 | | Hadean | |

*Time Millions of Years*

*The Geological Time Scale.*

time since then is called the Phanerozoic. That in turn is subdivided into three great eras, named Paleozoic, Mesozoic, and Cenozoic, or old life, middle life, and new life. Each of these three is further subdivided into periods that correspond to particular events that make up the "greatest hits" of the time of animals. The Paleozoic Era includes, in ascending order, oldest to youngest, Devonian, Mississippian, Pennsylvanian (which in Europe are combined into a single period named the

Carboniferous because of all its coal deposits), and Permian. The Paleozoic is followed by the Mesozoic and starts with the Triassic, followed by the Jurassic, and ending with the Cretaceous. Finally, the Cenozoic is divided into two periods, called the Tertiary and Quarternary.

So what do we see in the GEOCARBSULF results? Again, from oldest to youngest, the model estimates that there were lower than present-day oxygen levels in the early Paleozoic (Cambrian-Ordovician). Assuming that atmospheric pressure was the same as that today, which seems to be a reasonable assumption, according to climatologists, this would be equivalent to between 15-16 percent oxygen, compared to 21 percent present today.

The GEOCARBSULF results then indicate that there was a significant rise in oxygen, peaking about 410 million years ago, followed by a fall in oxygen levels in the middle to later part of the Permian. Mesozoic levels were also different from those of today, with GEOCARBSULF predicting lower than current oxygen levels, gradually rising to present-day values in the latter part of the Mesozoic.

It thus seems that Cambrian values were lower than present day and that the Permian witnessed the greatest drop in oxygen in Earth's history.

## CARBON DIOXIDE THROUGH TIME

While GEOCARBSULF has given us a new view of oxygen through time, a second pertinent question concerns carbon dioxide. The level of this gas has also varied through time, and, like oxygen, it has had important and far-reaching effects on the biosphere and on evolution. The main effect produced by varying carbon dioxide comes from its well-known greenhouse effects: during periods of relatively higher carbon dioxide levels, Earth will be warmer than during lower level times, as explained several pages above. Carbon dioxide through the last 600 million years has been modeled by GEOCARB III, and the most recent results, including characteristic climate for different times in Earth history, can be seen in the figure beginning the chapter, where R is the ratio of the mass of carbon dioxide to that of the present.

## EVOLUTION AND OXYGEN

Chapter 1 posed the iconic question of what determined the body plans that we observe in the fossil record and on Earth today. Evolution has created each of the body plans, but what specific factors were involved? Today, we can observe natural selection in action among microbes and among organisms with short generation times. Organisms are adapted to their environments and undergo evolutionary change if changes in their environment, either physical or biotic, affect their survival rates. Increasing temperature or pH are examples of physical environmental changes, while examples of biotic changes might be increased, or a new source of, competition or increased, or a new source of, predation. Evolution can occur in response to new opportunities and resources or as a defense against some new and deleterious condition. If we are to understand why the first animals recorded in the fossil record during the Cambrian period evolved the shapes and morphology that they did, we will have to have an understanding not only of the physical and biological characteristics of their Cambrian environments but of the changes to those environments as well. The history of life subsequent to the Cambrian Explosion can likewise be understood if we have a reasonable understanding of the physical and environmental changes that have occurred since the Cambrian. Evolution of animals can thus be understood as being caused by two different effects: modernization, where body plans increase in fitness through increases in morphological efficiency, and changes in response to environmental change, either physical or biological.

What have been the environmental factors that have changed the most since the Cambrian, thus stimulating evolutionary change? Most stocks of organisms have increased their efficiency of design over time and have found new ways of utilizing new resources, such as the vast stock of plant material on land following the evolution of land plants, and have responded to new types of predation, such as the Mesozoic Marine Revolution, proposed some years ago by University of California biologist Gary Vermeij. His hypothesis supposes that there was a great increase in predation during the Mesozoic Era, compared to the Paleozoic Era, brought about by the evolution of shell-breaking or -boring adaptations in many separate groups of animals. These adaptations caused a resultant evolution toward more shell defense in the prey organisms. But of the various nonbiological factors affecting life, changing oxygen levels were highly significant and the only factor that was global rather than regional.

In addition to changing oxygen levels, the significant nonbiological factors affecting evolution would include global temperature, the chemical makeup of seawater (such as pH and salinity), and the amount of sunlight hitting Earth. Of these, the swings in oxygen levels have been relatively greater than the changes in the others. While global temperatures have swung from relatively hothouse conditions during the Cambrian and Triassic through Eocene to glacial conditions during the Ordovician, early Permian, and Pleistocene, in reality there was never a time where some part of Earth, even during the most extreme temperature swings, did not maintain temperatures not only suitable but also favorable for animal life. Unlike the Snowball Earth (when the planet seemed to have undergone a very cold episode, perhaps freezing the oceans), episodes of 2.7 billion and 0.6 billion years ago did not involve pole-to-pole ice cover; there were still tropics at low latitudes.

In similar fashion, the hottest periods were never so hot as to threaten the existence of animal life. Swings in ocean salinity and pH have been even relatively less extreme. It is only oxygen content that has been both a global phenomenon and a parameter undergoing swings wide enough to affect not only life but also the evolutionary history of life. This factor perhaps has been rivaled only by biotic interactions of competition and predation in producing the makeup of animal body plans and their changes through time.

## A TEST OF THE HYPOTHESIS

The major hypothesis of this book is that changing atmospheric oxygen levels over the last 600 million years have caused significant evolutionary changes in animals. There has obviously been great change over this period of time, but are there times of greater and lesser evolutionary change and, if so, can these be correlated with oxygen levels? Restated, can the hypothesis that changes in atmospheric oxygen content over time have spurred evolutionary development (to form new species) and changes in body plan (or, that is, new kinds of morphology compared to the ancestral case) be tested with real data? The answers here are *yes* and *yes*.

In a 2002 article in the *Proceedings of the National Academy of Sciences*, James Cornette and colleagues Bruce Lieberman and Robert Goldstein showed a remarkable correlation between atmospheric carbon dioxide levels and rates of generic diversification (a measurement of new species formation: as we saw earlier a genus is the taxonomic rank above species) of marine animals. They took the extensive database records of species origination and extinction compiled by the late Jack Sepkoski and correlated these rates with atmospheric carbon dioxide levels as computed by Robert Berner and others. (It was in this earlier work that Sepkoski demonstrated that the rate of evolutionary change among animals, as measured by either the rate of new species origination or the rate of species extinction, has not been constant over the last 600 million years but has fluctuated.) Cornette and his colleagues found that the high levels of new species formation occurred in the early Paleozoic, most importantly during the Cambrian Explosion, which, it turns out, was a time with high levels of carbon dioxide. But they noticed that at other times with high carbon dioxide the data also showed high rates of new species formation. It seemed to Cornette and his coinvestigators that, somehow, high levels of carbon dioxide in the atmosphere triggered an increase in the rate of new species formation. But why? Animals do not use carbon dioxide in any way—just the opposite. Thus it is very puzzling to see these results.

Cornette and his colleagues explained their observations in the following way:

> The simplest hypothesis is that macroevolution is directly affected by carbon dioxide levels. Alternatively, paleotemperature may be an intermediary between the two systems.

But the first of these seems improbable; carbon dioxide, even at its highest levels since the evolution of animals in the Cambrian Explosion, was still at such minute concentrations that it was biologically neutral to animal life (although certainly it affected plant life, since higher carbon dioxide stimulates more growth—but not necessarily more new species), and even if it were at higher levels, animals do not use it for any aspect of living. But the second part of their statement is more plausible. This latter statement links the known correlation between carbon dioxide levels and planetary temperature; because carbon dioxide is such an efficient "greenhouse gas," when the level of carbon dioxide rises in the atmosphere, the planet warms. Thus, Cornette and his team suggested that temperature change has been the most important factor. They stated:

> One might even hypothesize that high temperatures directly increase marine diversifications or that low temperature and specifically glaciations inhibit marine diversification. . . . Additionally one might pose a hypothesis that some factors that enhanced plant diversification inhibited marine diversification. . . . Yet another hypothesis is that enhanced carbon dioxide levels may be associated with increased seafloor spreading rates that could encourage biological diversification.

Thus, the three possibilities advanced by Cornette et al. are that the rate of new species formation rises when carbon dioxide levels are high because (1) high levels of carbon dioxide somehow cause new marine animals species to form; (2) higher temperatures somehow cause new animals species to form; or (3) when plant diversification slows during times of dropping carbon dioxide, so too, somehow, does marine animal diversification. It is evident that during much of the past 400 million years, periods with high carbon dioxide were times of low oxygen and that the reverse was true as well. It is only for the Cambrian and early Ordovician that this inverse relationship does not seem to hold. Thus, the alternative to the Cornette et al. hypothesis is that it was *not* high carbon dioxide (and thus warm temperatures) that stimulated high speciation rates, but low oxygen. It turns out that by doing a statistical test comparing oxygen vs. species formation rates, a highly significant correlation can be found.

These results are striking. Carbon dioxide values may have had little or nothing to do with the changes in animal diversification rates. Instead, it can be proposed that oxygen levels, or perhaps the rapid change of oxygen levels from those less than somewhere around 15 percent by volume to present-day levels, stimulated new speciation rates, in response to animals attempting to cope with a reduction (or already low level) of oxygen in the early Paleozoic, but high levels of oxygen are evolutionarily neutral. This is because animals that have

adapted for low levels of oxygen become more efficient in higher levels and are not required to make any morphological or physiological
changes. But animals that evolved in high levels of oxygen are severely
affected when oxygen drops. They must make major morphological
and/or physiological changes or become extinct. Niles Eldredge
and Steve Gould, more than three decades ago, in their now famous
Theory of Punctuated Equilibrium, demonstrated that most morphological changes occur during speciation events. Most morphological
changes happen quickly when a new species forms, not gradually.
The changes necessary to adapt to low oxygen levels (here somewhat
arbitrarily chosen as <15 percent, based on the results of the correlations described above) involve morphological adaptation in existing
lineages (such as size decrease) but more probably required such extreme morphological change that a new species would be created, incorporating new adaptations to lowered oxygen. Fluctuating oxygen
levels, rather than carbon dioxide levels, proposed here are a more
likely explanation for the interesting periods of higher and lower evolutionary change shown by marine animals.

If periods of low (or lowering) oxygen are times when the rate of
new species formation is high, we might expect that these times would
also show high diversity—that is, the number of taxa present at a given
time would be high. This can be examined by plotting oxygen concentrations through time against diversity through time, and again there
is a significant correlation. These results are being published in scientific literature.

A terrestrial vertebrate data set is also available. Once again, if we
plot atmospheric oxygen over time, we see the same relationship. By
far the greatest origination rates for land animals during the Paleozoic
occurred after the Permian extinction. This high peak is in response to
the elimination of most land life tetrapods (four-legged animals had to
virtually start anew, thus stimulating a very high origination rate). A
more telling finding is just before this peak: the late Paleozoic interval
of high oxygen shows a low rate of origination on land. *Thus, in the sea
and on land, the time of high oxygen was a time of stagnant evolution.*
Beyond affecting diversification rates, how else might oxygen levels
have affected land life? I propose that terrestrial vertebrates can be
roughly divided into two large assemblages. One group evolved and

flourished in high-oxygen conditions, the other in low-oxygen. With this perspective, changing oxygen conditions led to ecological replacements of one group by another. For example, a large pattern of diversification throughout the Triassic (a time of dropping oxygen to low levels) was the result of high-oxygen terrestrial vertebrates, such as the mammal-like reptiles (the group that was the direct ancestor of we mammals, an event taking place in the Triassic Period of the Mesozoic Era) being replaced by low-oxygen organisms, most importantly the saurischian dinosaurs, which were the first-ever dinosaurs.

Let's summarize this section. While it cannot yet be demonstrated that it was the *change* in oxygen that actually stimulated the evolutionary change (correlation does not imply causation), our understanding of the importance of respiration to all animals leads to the conjecture that the change in oxygen values was indeed the major stimulus. This can be formalized as follows:

*Hypothesis 2.1: Reduced levels of oxygen stimulate higher rates of disparity (the diversity of body plans) than do high levels of oxygen.*

*Hypothesis 2.2: The diversity of animals is correlated with oxygen levels. The highest diversities are present during times of relatively high-oxygen content.*

For terrestrial vertebrates, oxygen levels of less than about 15 percent seem to promote the formation of new taxa, stimulated largely by the anatomical and perhaps physiological needs of organisms in lower-oxygen environments. Another aspect of the influence of low-oxygen levels on animal disparity and diversity is that times of lowered oxygen also produced increased partitioning of land surface by what Ray Huey and I, in a 2005 paper in *Science,* have named "altitudinal compression." Our hypothesis suggests that during times in Earth's history of lowered-oxygen values, even modest elevations would have become effective barriers to gene flow and thus would have stimulated new species formation by isolating populations. Thus, there would have been more endemism (animals found in small geographic areas) during the low-oxygen times.

Why, then, would there be a difference in evolutionary rates between low-oxygen and high-oxygen times? Very simply, animals that evolved in low oxygen not only survive but also in some cases thrive in high oxygen (birds are a notable example). In other cases, however, the use of low-oxygen lung systems no longer needed in high-oxygen conditions may have caused some competitive replacement and thus stimulated new evolution. The replacement of many saurischian dinosaurs by the group known as ornithischians, or "bird-hipped dinosaurs," and mammals in the Cretaceous may be an example of this as well. But, more significantly, times of low oxygen seem to have been major intervals of evolutionary innovation, leading not only to new species but also to new kinds of body parts, such as larger or even new kinds of lungs. Perhaps it is more correct to posit that periods of lower oxygen served to foster increases in disparity (the measure of morphological rather than species diversity) and diversity of species.

## MASS EXTINCTIONS AND LOW OXYGEN

A final aspect of oxygen's effect on the history of life must be noted. As we will see in the pages to come, the history of life was punctuated by a series of mass deaths (at least 15 over the past 500 million years) when significant numbers of animals and plants rapidly went extinct. Five of these involved the extinction of more than half the world's fossil-producing species and were named "The Big Five" by extinction specialists David Raup and Jack Sepkoski in the 1980s. At the time of four of these five mass extinctions there was a common environmental condition. All occurred either in a time of very low oxygen (generally <15 percent, as was the case for the Ordovician, Devonian, and Triassic mass extinctions) or happened after at least a 10 percent drop in oxygen (as was the case for the Permian mass extinction). And it was not just the major mass extinctions. Even relatively minor drops in oxygen level were coincident with wholesale species disappearances. In most cases of low oxygen there was high carbon dioxide and thus high temperature and physiologists have repeatedly observed that higher temperatures increase the stress of low oxygen on aerobic organisms. Exactly why this happens is not well understood, but happen it does.

Without exception, these mass extinctions were associated with a dip in oxygen levels. It may not be the low oxygen per se but the change in the level of oxygen that caused the extinctions.

## BACK TO THE PAST

With this background on animal respiration and the geological oxygen record, we are now ready to tackle the history of animal life as recorded in the rock record. The next chapters will examine in more detail major episodes in the history of animal life, their relationship to oxygen levels, and morphological adaptations that appear to have been brought about in response to changing oxygen. Thus, each chapter will not only deal with diversification and extinction trends but will also feature as well new hypotheses on specific morphological adaptations that appear to be related to oxygen levels, which can be viewed as the evolutionary attempts of various animals to maximize respiratory efficiency.

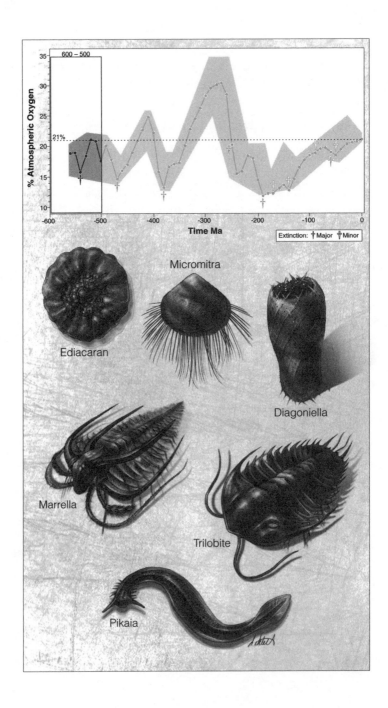

## 3

## EVOLVING RESPIRATORY SYSTEMS AS A
## CAUSE OF THE CAMBRIAN EXPLOSION

The origin of animals is the start of this story. Its timing has been hugely controversial, with two distinct lines of evidence giving quite different views on the timing of the first diversification of animal phyla, that seminal period when some first animal ancestor evolved into many kinds of animals. One of these lines comes from the pattern of appearance of animal fossils in rocks, the second from molecular clock studies on extant animals. This chapter will look at these records for clues to one of the greatest of all paleontological mysteries: what caused the rapid diversification of animal body plans in the Cambrian Explosion?

When did the first animals diversify into the numerous animal phyla? Since the time of Darwin it has been known that fossils of animal life seem to appear suddenly in the fossil record. Over short stratal intervals, sedimentary rocks seemingly bereft of fossils were found to be overlain by rocks with a profusion of highly visible fossils, the most common being trilobites. Trilobites are fossil arthropods: the familiar animals such as insects and crustaceans are unified into a single phylum through the shared presence of hard, jointed exoskeletons, and trilobite fossils are the remains of highly evolved and complex animals. This observation was highly vexing to Darwin (and hugely comforting to his critics), as it seemed to fly in the face of the then newly proposed

theory of evolution. Today, we know that animal life did appear com-
paratively rapidly in the fossil record, and new dating techniques now
put the time of the first complex fossils at slightly older than 540 million years ago, with the first trilobites appearing in the record some
single-digit millions of years after that.

In contrast to this slightly more than half-billion-year-old age for
the first animals, molecular studies of extant animals looking at the
age of divergence of basic animal lineages suggested a far older origin
of animals. One influential study by Charles Wray and his associates at
the American Museum of Natural History dated the divergence of animals from protozoan (single-celled organisms, as compared to animals, which are all composed of many cells) ancestors at about a billion
years ago. In this latter view, animal phyla diverged early but remained
at a very small size and invisible to the fossil record for a half billion
years. Both views agreed that the appearance of animals in the fossil
record was a significant event, which has been called the Cambrian
Explosion. To paleontologists, the Cambrian Explosion marked the
first evolution of animals. To molecular geneticists, it marked the first
evolution of animals large enough to leave remains in the rock record.
The controversy raged through the 1990s, to be resolved in the early
years of the twenty-first century when new molecular studies, using
more sophisticated analyses, essentially confirmed the younger date
for the origin of animals that had been championed by paleontologists. There is now agreement that animal life on Earth did not predate
600 million years ago and might be closer to 550 million years in age.

The Cambrian Period is now dated from 544 million to about 495
million years ago. Over those roughly 50 million years, the vast majority of animal phyla first appeared. All specialists agree that this is the
most important event in the entire history of animal life, superceded
in importance only by the first appearance of life on Earth, perhaps, in
the context of the entire history of life on our planet.

The Cambrian Explosion has thus been a "hot" topic in science. A
vast library of books, technical articles, and popular accounts of the
Cambrian Explosion exists. The paper devoted to these many pages
has probably destroyed whole forests, which is a bit ironic, since the
Cambrian period occurred long before vascular plants colonized the

land, let alone evolved into trees and forests. With so much written by so many, what more can be said? With new and better estimates of the oxygen levels back then, the topic is ripe for reexamination.

## OXYGEN AND THE CAMBRIAN EXPLOSION

The levels of carbon dioxide and oxygen postulated for the Cambrian Explosion interval are shown in the figure beginning this chapter, taken from the Berner curve introduced in the preceding chapter. According to this curve, oxygen levels soon after the start of the Cambrian Explosion 544 million years ago were about 13 percent (compared to 21 percent today) but then fluctuated. During this time carbon dioxide levels were far higher than they are today, tens of times higher in fact, and such high levels would have produced a greenhouse effect. Even with the drop in carbon dioxide levels at the end of the Cambrian around 495 million years ago, temperatures at this time would have been higher than in the present-day. Since less oxygen is dissolved in seawater with higher temperature, the already anoxic conditions of the oceans, due to the low atmospheric oxygen of that time, would have been exacerbated.

## A CAMBRIAN DIVE

To explain the conditions, flora, and fauna of the Cambrian, imagine that we have journeyed back in time to Earth circa 522 million years ago. To really do this right we need not only a time machine but also a spaceship, in order to better see the position of the continents, for continental position and geological processes accruing from plate tectonic processes had a determining effect on subsequent biotic history.

The first thing we notice as we pass over the land surface is that there is so little vegetation. Low traces of green can be seen in the wetter areas but most of the land surface is bare rock. It looks like the areas around glaciers at high altitudes in our present-day world, but even as we cross the equatorial regions the starkness of the place is apparent. Plant life is limited to moss and vast slicks of plant-like, photosynthesizing bacteria. There is no rich organic soil. There are no trees, no bushes, no flowers at all. There are no vast grasslands, nor are there any animals.

Even the nature of the river systems seems bizarre. While huge rivers do cascade out of the mountains, as they make their way toward the coastline of the Cambrian Ocean, they remain low and anastomostic, a series of braided streams passing over the gravel of the land surface. Nowhere do we see the giant meandering rivers of our time. There are no lazy river bends, no oxbow lakes, no point bars that future fishermen will stand on casting for trout. There are not even trout, for that matter, or any other visible creature at all in these rivers. No bugs, no minnows, no amphibians, no wading birds, no dragonflies, or anything else so familiar to those who revere and visit the grand rivers of the Planet Earth we know today. The great mountains are being eroded, as they will be through time, but the sedimentary refuse resulting from erosion is passed to the sea in rivers that have been the same now for the 4 billion years since Earth came into existence. That will soon change; riverbanks will, some tens of millions of years hence, become colonized by rooted plants, and rivers will begin to meander over vast floodplains and river valleys. That will come, but not on this day.

Our craft passes over the last of the coast plain and we see the sea. Here at last something looks familiar: waves and currents that look no different from those seen in our world. But once again, when we descend for a closer look, the alienness of this world becomes apparent. We park our craft on the seashore, step out, and gasp in air that is not the thin air of the high mountains but air that gives our lungs no traction. It is as if we have stepped out onto Mount Everest in terms of the amount of oxygen but the seashore in terms of air pressure. This is not thin air, it is low-oxygen air. The air pressure is the same as we experience at sea level in the present, but the amount of oxygen making up the air is lower. With our oxygen masks in place, we walk the sandy shore. The sand is composed entirely of minerals. There are no flakes of seashells, bits of coral, or skeletons of planktonic creatures like foraminifers or coccoliths. It is silica-rich sand bereft of organically precipitated content.

The beach is huge. We walk from the waves' edge to the high-tide land, marked still by foam and the wet mark of the recent high tide. The tidal change here is at least 15 meters of vertical distance between the high-tide line and the ocean's level, a tidal change that can be found only in the Bay of Fundy, Canada, in our own world. We look up at a noticeably dimmer sun, dimmer because all stars grow more energetic through time, and to the east we see a half moon just risen, dim as well in the afternoon sunlight but far greater in size than it will be in our time. The moon is clearly closer to Earth in this long ago time, and that would explain the great tidal range.

We walk to an outcrop of hard basalt that makes one of the headlands of this broad sandy beach, and even in areas of the outcrop where the low tide still washes up, there are few of the familiar features of our world. There is no upper tide assemblage of *Littorina* (the common seashore periwinkle snails), or barnacles beneath the snails, or lines of mussels beneath the barnacles—no intertidal zonation at all, in fact. But it is not dead, this rock. There is a variety of algae clinging to it, reds and browns that wash in the rushing current, not all that different from the kelps of our world, except for the fact that no animals live amid them.

It is time to look at what lives in the sea. This being a thought experiment, we suit up with appropriate diving gear, clean our masks, don flippers, and dive downward. We are not the first to take this trip. In his 1998 book, *The Crucible of Creation*, Simon Conway Morris took readers on a submersible trip into the Cambrian Ocean at the site of what would become the Burgess Shale and in so doing described the vast assemblage of organisms that would become the fossil finds of that most important of all fossil deposits, our best view into the world at the height of the Cambrian Explosion. But the Burgess Shale was deposited some 510 million years ago. Here we are in the world of 522 million years ago, almost 12 million years before the Burgess. Twelve million years of evolution is a long time for animals to diversify. We are at the beginning of things, and the place we dive into will come to be known as Chengjiang, China, and, like the Burgess Shale (if not so famously), it will yield fossils with soft parts, giving us a window into the start of the Cambrian Explosion.

We descend through the surf zone and make our way offshore into deeper water, and, on a muddy bottom still somewhat influenced by wave action at lowest tide, we scan the bottom for life. And life there is—in abundance. The first noticeable organisms are numerous small sponges, most attached to cobbles on the bottom (but with a few rolling slightly with the slight current on the bottom, among the small ripples found in the bottom sediment). A quick census shows at least 30 different kinds, probably each a separate species. Most are the familiar demosponges of our time, but a few hexactinellid, or glass sponges, are found in this undersea forest of sponges. Other phyla are seen as well. There are a few anemone-like creatures, and floating with us in the water column are jellyfish and some ctenophores.

We descend onto the bottom itself and sift through the dark sediment. Very quickly we find a diversity of small worm-like forms. Most are priapulids and one, which will eventually be named *Maotionshania*, is particularly abundant. We search further and find buried clam-like creatures that on closer inspection are found to be inarticulate brachiopods looking much like the still extant *Lingula*. These have long and mobile tethers, pedicles that help them dig back into the sediment when we release them back onto the sea bottom. Other inarticulate brachiopods are found encrusted on rocks. Among them are small numbers of another shelled invertebrate, the tube-shelled hyoliths, with their strange arms called "Helens" in honor of the daughter of their discoverer, Charles Wolcott.

Small worms, sponges, bivalve brachiopods, and the tiny hyoliths, these are the minor elements of the fauna. The rest of the fauna here belongs to one phylum, and it is present in a staggering diversity. It is the arthropods that dominate this sea bottom, totally eclipsing the other invertebrates in all of number, diversity, and size. This is truly a world of segmented "bugs."

They are everywhere. The most common of all are small, ostracod-like bradoriids and tiny bivalved arthropods about 1 millimeter long. There are lots of other bigger arthropods. Some, such as *Naraoia*, that will survive into Burgess time show little or no exoskeleton. As we approach, some hurriedly roll up like pillbugs for protection. Other arthropods that will be found in the Burgess Shale are here as well, including the bivalved *Canadapsis*. Large forms with stalked eyes, to be called *Vetulicola* and *Chuandianella,* watch us carefully, while another form with a short head shield, bulbous eyes, and large spines on its tail region scuttles away, crab-like. Amid these strange arthropods are the more familiar trilobites, but, unlike in later Cambrian deposits, they are fewer in number and diversity than their more esoteric cousins. Most are the familiar redlichiacean trilobites with numerous segments, forms that Darwin and his contemporaries considered the oldest animals on Earth and forms that are so striking because of their many segments. The trilobites here are heavily armored compared to the many other thinner carapaced arthropods among them, and these other poorly skeletonized forms will not commonly fossilize. At all localities save this one and a few others, which because of special circumstances will preserve soft parts, only the trilobites will be found, giving the false impression that they were the most common members of the fauna. Here we see that they are only a minor part of the fauna.

While most of this arthropodan assemblage is fairly small in size, we see bigger animals too. There are several arthropod forms that look like sea scorpions and some that have flattened oval-shaped bodies such as the enigmatic *Saperion.* All of these are somewhat intermediate in size, and now we search for the top carnivore of this ecosystem—it too is an arthropod—and we do not have long to wait. Swimming lazily through the water, some meters above the teeming bottom, we see a meter-long *Anomalocaris,* famous from the Burgess and here as well, showing its antiquity. It settles downward onto the bottom, its large paddlelike tail slowing as it lands with its many walking legs taking up the shock. With large claws slashing, it begins to feed on the many smaller, fleeing arthropods. The *Anomalocaris* takes note of another not as big but still substantial invertebrate on the bottom, a heavily armored lobopod, also an arthropod but one that is very rare today. This strange animal looks like a cross between an annelid worm and an arthropod and seems to be related to the still-living onycophorans of our world. The phosphate plates on the lobopods provide some protection but soon it too is killed, and the *Anomalocaris* centers itself over the body and begins feeding with its peculiar, circularly plated mouth.

It is time to go. As we head back toward the seashore we see one more animal. It is not an arthropod, but it does have a segmented side.

It is a primitive fish-like creature, one of several that live here. Some look like eels, some like hagfish. They are the first chordates, or vertebrates. They are our ancestors.

## CAMBRIAN LIFE

The distribution of fossils from the fantastic deposits in Chengjiang, China, has given us a new window into the origin of the animal phyla on Earth. To continue the analogy, this is a window to a floor lower than that of the Burgess Shale. The approximately 12 million years separating the age of these two deposits thus gives us a new view of how animals diversified. Because both Chengjiang and the Burgess preserve soft parts and skeletonized animals, we have a good picture of what was there and in what relative abundance. Without this added view yielded by the preservation of soft parts, we would never be sure about the relative abundance of various kinds of animals, for perhaps there was a huge abundance of creatures like soft worms and jellyfish, forms that had no skeletons. Thus we are surprised at what appears to be a clear view of the nature of the fauna at both sites. Over 50,000 fossils have been collected from the Burgess Shale (and a lesser number from Chengjiang). In their summary of the Burgess fauna, Derek Briggs, Doug Erwin, and Fred Collier, in their 1994 book *The Fossils of the Burgess Shale*, list a total of 150 species of animals. Almost half are arthropods or arthropod-like. But an even more interesting number relates to the number of individual fossils. Well over 90 percent of all Burgess fossils are arthropods, followed in number by many fewer sponges and brachiopods. Like the earlier Chengjiang, the Burgess sea bottom was dominated in the kinds and numbers of animals by arthropods. Arthropods are among the most complex of all invertebrates and yet in these almost earliest of fossil deposits in the time of animals they are diversified and common.

## THE PUZZLE OF THE TRILOBITES AND
## THE ORIGIN OF SEGMENTATION

Our visit back to the Cambrian leads to an inescapable conclusion: in sheer numbers of individuals and species (described as diversity) and in sheer numbers of different kinds of body plans (described as disparity), the arthropods were the most successful of Cambrian animals. How much of this success was due to their principal body plan characteristic—segmentation?

Segmented animals are the most diverse of all animals on the planet today, and most are arthropods. All arthropods, including the highly diverse insects, show repeated body units and body regions based on groupings of individual segments that have specific functions for the animal. The feature uniting the group is the presence of a jointed exoskeleton that encloses the entire body. This exoskeleton even extends into the gut. The exoskeleton cannot grow, so it must be periodically molted and another, slightly larger, one produced. The body has a well-differentiated head, trunk, and posterior region in varying proportions. Appendages are commonly specialized. On terrestrial arthropods the appendages are usually single and enormous, but the marine forms generally have two branches or parts per appendage, an inner leg branch and an outer gill branch, and are thus termed "biramous."

Arthropods are not alone in being segmented. All annelids are segmented, and some members of generally nonsegmented groups, such as the monoplacophoran mollusk, show segmentation. It appeared early in the history of animals and indeed the Cambrian trilobite fossils, the most common of the earliest preserved animal, show segmentation.

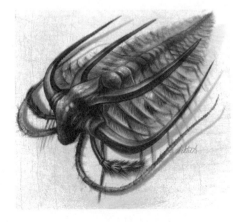

*Reconstruction of* Marella, *a Cambrian arthropod from the Burgess Shale. As can be seen here, each segment behind the head bears a pair of large gills. The area of gills to overall body volume is thus very large in this kind of body plan.*

The exoskeleton encloses the soft parts like a suit of armor and that may be its major function: protection. But the consequences of this kind of skeleton are huge: there can be no passive diffusion of oxygen across any part of the body. To obtain oxygen the first arthropods, all marine, had to evolve specialized respiratory structures or gills. The combination of segmentation and an exoskeleton used by the arthropods was clearly a design winner: the arthropods, all with this design, have more species today on Earth than any other phylum. Some (perhaps most) of that success must be due to their characteristic segmented body. In his 2004 book, *The Origin of Phyla*, James Valentine reflected on what is a major evolutionary puzzle: why were there so many kinds of arthropods in the Cambrian and such large populations of individuals belonging to the many kinds of this group then present? It is worthwhile to look at what he has written on this subject:

> A marvelous diversity of early arthropod body types has come to light, so many and so distinctive as to pose important problems in applying the principles of systematics. The most diverse of the extinct arthropod groups is the Trilobita. . . . However, a large number of non-trilobite fossils with jointed bodies and appendages display great disparity in just those features that form the defining characteristics of the living higher arthropod taxa— tagmosis, including segment numbers and the number, type and placement of appendages. Most Early and Middle Cambrian forms have such unique assemblages of these characters [body parts] that they cannot be included in any of the living higher taxa as they are defined within crown groups and many of the fossil taxa are quite distinct from each other as well. These disparate arthropod types are phylogenetically puzzling. . . . This evidently sudden burst of evolution of arthropod-like body types is outstanding even among the Cambrian Explosion taxa.

Hence we have an interesting puzzle. What we call arthropods are composed of what appear to be many separately evolving groups that have, through convergent evolution, produced body plans of great diversity save for one aspect: all have limbs on each segment that are biramous—each appendage carries a leg of some sort and a second appendage, a long gill.

Why would early primitive (or basal) animal groups opt for segmentation? Perhaps this is the wrong word, for Valentine and others have noted that the arthropods are not so much segmented—which at least in annelids consists of largely separated chambers for each segment of the body—as *repeated.* Valentine proposes that this striking body plan arose in response to locomotory needs, stating:

> Clearly, the segmented nature of the arthropod body is related to the mechanics of body movement, particularly to locomotion, with nerve and blood supplies in support.

There is no doubt that this type of body plan is an adaptation that aids locomotion. But here we can depart from Valentine and suggest that the main function of this kind of body plan is to allow repeated gill segments, each small enough to be held in optimal orientation beneath the segments. The flow of water across these gills, while at first glance appearing to be passive, may actually represent a pumping gill.

Look at trilobite morphology. While the upper surfaces of the trilobite carapace is commonly preserved, the underside, bearing appendages and gills, is rarely preserved.

Trilobites have long been known to have a curious food acquisition system. The same appendage terminations mark the fusion of the walking legs and gills and end in a blunt paddle structure known as a gnathobase. Paleontologists have long surmised that a forward-moving current created by limb movement would move food material to the mouth underneath the animal. But such a current would also serve to bring new water across the gills, and with the carapace of the animal sitting overhead like a roof or tent, the trilobite or other arthropod could build a directed water current defined by the body above and the sediment or sea bottom below. I propose that oxygen acquisition purposes primarily led to this body plan and that it was only secondarily co-opted for food acquisition. The environmental conditions of oxygen levels far lower than those of today would have provided the stimulus to begin this evolutionary pathway leading to the arthropod body plan.

*Hypothesis 3.1: The repeated-segment body plan came about to allow the formation of a large gill surface area, with the gnathobase water current system evolving as part of this respiratory structure. The overall shape of the arthropod creates a defacto pump gill system. Segmentation evolved as a way of increasing gill surface area during the latest Neoproterozoic (from about 600 million years ago to 538 million years ago).*

So why is segmentation needed for a large gill surface? The same amount of gill surface area could be evolved and exist as one or a small number of large sheets attached to the body. But such large structures would easily fold on themselves and could also become problems during swimming by increasing drag.

How did this kind of repeated body part system come about at all? I propose that an important group of regulatory genes known as Hox genes were themselves co-opted in order to provide repeated gills.

A rich new topic of research spanning the last two decades has been at the interface of development and evolution. There are rules to the formation and growth of embryos as the genetic code of the genome becomes expressed as the protein and protoplasm of the growing individual. Newly discovered "regulatory" gene complexes, such as one called the Hox gene complex, are increasingly viewed as being nearly as important as the genetic code itself. As for segmentation, biologist S. A. Newman has postulated that it is a consequence of developmental regulation and pattern, not a feature evolved specifically for function. In other words, segments initially may have come about not because they work better in the day-to-day life of an organism with them but because they made growth and the many morphological changes occurring in these animals from embryo to adult take place more smoothly. Once in place, however, the presence of segments was molded by evolution for use in specific functions such as locomotion, respiration, feeding, and reproduction through evolutionary processes placing specific locomotory, respiratory, and reproductive appendages on various segments.

Do animals with segments really have an advantage in respiration over those that are nonsegmented? Let's look at a measure of comparing respiratory structures, starting with passive systems. In a passive system, as mentioned in Chapter 1, the oxygen-carrying medium, in this case seawater, comes in contact with the respiratory epithelium. Oxygen diffuses across the membrane into the circulatory fluid and is moved to the various parts of the body and its cargo of cells, all needing oxygen. Carbon dioxide is exchanged across the same membrane out of the body. Three factors control the rate at which oxygen can be extracted: its concentration in the water, the rate at which it can be diffused over the membrane (largely a function of the thickness of the membrane), and the volume of new water that comes in contact with the gill surface (largely a function of the size of the gill and the rate of water movement over the gill surface).

The simplest way to increase the volume of new water in contact with the gill surface is to increase the size of the gill, and one way to compare oxygen-extraction efficiencies is to compare relative gill sizes for a given volume of the animal whose gill it is. Thus, a simpleminded way of comparing invertebrates in terms of oxygen-extraction efficiency is to measure gill surface and animal volume and then compare them, assuming constant thickness of the respiratory surfaces. In other words, we are assuming that all gills are equally efficient for unit area, an assumption that is probably not completely correct but close enough to allow an adequate comparison.

Trilobites turn out to have a very high gill to body volume ratio, based on as yet unpublished observations made in the course of writing this book, compared to most modern-day crustaceans, especially all crabs. We really cannot determine whether trilobites and other segmented arthropods have a respiratory advantage over other, nonsegmented animals, although clearly their segmented body plan has been extremely successful. Let's also look at the other major Cambrian animals and their respiratory successes.

## RESPIRATORY SYSTEMS OF OTHER CAMBRIAN ANIMALS

Many of the animals present in the Cambrian had no specialized respiratory structures, and they generally shared a common characteristic: they were covered with epithelium that allowed for the passage of water directly from the sea into the body of the animal. For them the entire body was a gill. Examples include many soft worm-like forms, such as nemerteans and the many varieties of the jellyfish-sea anemone clan, the cnidarians. The cnidarians were and are composed of but two cell layers, and they all show a gut cavity that also allows respiration. Adding to this basic structure of the cnidarians are many tentacles and in larger forms, such as the large sea anemones, there is an interior gut that shows complex invaginations for digestive purposes. Thus, these animals have a very high surface area to volume ratio. Cnidarians prove themselves to be characterized by the highest surface area to volume ratios of any animal yet examined. While there are few fossils of cnidarians from the Burgess Shale or Chengjiang deposits, those that we have (such as the numerous examples of pennatulaceans, or Sea Pens) demonstrate the high surface to volume ratio of this phylum.

Sponges are another animal found in abundance in the early Cambrian deposits. Like the cnidarians, sponges show no respiratory structures, nor would we expect any. With a body plan built around a series of sacs (like the cnidarians but with even less organization: there are no true tissues in a sponge), all sponges show a very high surface area to volume ratio. In fact, sponges are like agglomerations of numerous, single-celled organisms, with each cell essentially in contact with seawater. But even with this advantage, sponges show an even more efficient way of gaining oxygen. Their main feeding cell, called a choanocyte, causes large volumes of water to pass through the sponge. Some sponge specialists have suggested that a sponge passes as much as 10,000 times its volume in seawater through its body each day. This puts the sponge in the category of having functional "pump gills," since the feeding cell forces water past the cells at a phenomenal rate. Consequently, sponges are capable of living in extremely low-oxygen conditions.

So if most soft-part-covered animals, except for the very large ones, do not seem to require specialized respiratory structures, what of those with hard parts? As noted earlier in this chapter, the major groups of animals with hard parts in the Cambrian were the huge tribe of arthropods, followed in numerical importance (in most Cambrian marine strata) by brachiopods and then by a smaller number yet of mollusks and echinoderms. We have already discussed the respiratory system of arthropods, particularly trilobites, and will discuss these less numerous groups next.

First, Cambrian echinoderms make up a weird assemblage of small boxlike animals. All echinoderms have an outer layer of living flesh. The oxygen needs of this layer are taken care of by direct diffusion of oxygen from seawater and by the layer's formation with gill-like elements emanating from the flesh in some modern forms, such as sea urchins, starfish, and sea cucumbers. All of these echinoderms also have thin-walled tube feet, used for locomotion but also serving for respiration, as oxygen can diffuse across the tube feet and then pass inward through small holes in the skeleton (the ambulacra) to the interior of the box. There were many curious forms related to the more familiar kinds in our world. All of these echinoderms surely were covered with a thin epidermis as well, so that only their interiors needed enhanced respiratory techniques. All had hard skeletons of outer plates that surely inhibited oxygen uptake, thus requiring some form of adaptation for respiration for the interior. This need for interior respiration is somewhat alleviated by the fact that most echinoderms have very little flesh in their boxes. A sea urchin, for instance, while having a voluminous space in its spherical skeleton, actually has little flesh within; our measurements of Puget Sound urchins show that only about 20 percent of the volume within their box is composed of living protoplasm. The rest is seawater or the unique water vascular system that enables the tube feet.

More common than echinoderms were mollusks. Most during the Cambrian were small in size. Each of the major classes of mollusks (gastropods, bivalves, and cephalopods) is found in Cambrian strata. The most common mollusks were monoplacophorans, a minor class today but common in the Cambrian. They had a limpet-like shell and a snail-like body with a broad, creeping foot. Most interestingly, alone among mollusks of the time, they showed a body organization that suggests segmentation. From looking at muscle scars on the fossil shells and from comparing anatomy of the still living forms, it looks as if the Cambrian monoplacophorans had multiple gills. Modern-day gastropods have a single pair of gills or sometimes a single gill. But the Cambrian monoplacophorans, which lived a very snail-like existence in all likelihood, found it necessary to have multiple gills. The respiratory adaptations of the mollusks are distinctive enough to warrant additional discussion, below.

Second in abundance to the arthropods were brachiopods, a phylum related to bryozoans that are routinely mistaken for bivalved mollusks. Yet while the shells of bivalves and brachiopods show a superficial similarity, the internal anatomies of the two groups are radically different. The major feature of a brachiopod is a feeding organ known as a

Micromitra, *a Burgess Shale bra-*
*chiopod. Although looking like a*
*clam, its internal anatomy is very*
*different from any mollusk.*
*There is a large organ called a lo-*
*phophore that, in shower-curtain*
*fashion, separates the interior of*
*the shell into two compartments.*
*As water is forced over this*
*net-like structure, both food and*
*oxygen are extracted. Thus, all*
*brachiopods are examples of*
*creatures using the pump-gill*
*respiratory system. The spines on*
*the outside of this species are not*
*part of the respiratory system.*

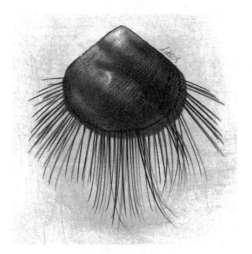

lophophore, composed of a large loop with numerous long, thin fingers producing a delicate fan-like shape in the shell. This organ filters
seawater for food, and, as it is filled with a body fluid and is very thin, it
also serves as an exquisite respiratory organ. Like the body of an echinoderm, the interior of a brachiopod shell is almost all water. There is
very little flesh, which is in contact with a steady flow of seawater.

The brachiopod lophophore creates several currents of seawater
that pass into the sides of the shell, move across the lophophore, and
are then sent out the front of the shell. This constant stream of new
water entering a brachiopod has the same effect as the current passing
through a sponge. The small volume of flesh to great surface area of
the lophophore, coupled with the steady flow of water (many times the
volume of the interior of the shell), makes the brachiopod consum-
mately adapted for a world of low-oxygen. While the lophophore is
explained as primarily an adaptation for feeding, its respiratory function may, in fact, have been a primary aspect of its evolution and shape
in the first place. By building a shell and then creating the water currents passing over the lophophore, the brachiopods appear to have
been among the first examples of an animal using a "pump gill."
Sponges also pump but they do not have an internal organ (nor indeed

do they need one) to deal with oxygen extraction. But the brachiopod shell design allows a clever combination of feeding and respiration in the same organ. This design appears to be a classic example of a body plan developed either partially or mainly to deal with life in a low-oxygen environment.

## EVOLUTION OF THE MOLLUSCAN BODY PLAN WITH EMPHASIS ON RESPIRATION/OXYGEN

Following arthropods, mollusks are the most diverse groups of animals on Earth today, and surely it was the same in past times. In terms of body plan they are very different from arthropods. Only a few groups of mollusks show any sort of body part repetitiveness; mollusks have a single unsegmented body, and none are jointed. The dorsal body wall is cloaked by a large flap of tissue known as the mantle, and this tissue secretes the calcium carbonate shells that are found in the majority of mollusks, while the ventral surface is primitively shaped as a broad, flat "foot" that allows locomotion. The mouth is anterior and the anus posterior, and most have an open circulatory system. There are exceptions even to this, however, the most important being the cephalopods, a group of mollusks that we will look at in detail below.

Mollusks are known from the early Cambrian, and they had probably already differentiated into many separate groups by the earliest Cambrian time. The group appears to have evolved from some type of flatworm. Early in the history of mollusks, however, the evolution of the dorsal shell created the same problem that was faced by arthropods: with a shell covering, large areas formerly used for passive oxygen uptake were no longer available. The standard explanation is that, like the arthropods, the evolutionary response to the new barrier to diffusion created by the shell was to create a gill system. Here we can propose a novel scenario: that the shell evolved to increase the efficiency of the respiratory process rather than solely as a protective organ. Like so many pathways in evolution, the mollusk shell evolution pathway clearly involved multiple uses of the evolved structure, and protection was a consequence of shell formation, protection not only from predators but also perhaps more importantly in the latest Precambrian protection from harmful ultraviolet radiation or desiccation in intertidal settings. This idea can be formalized here:

> *Hypothesis 3.2: Initial shell formation and subsequent shell geometry elaboration in classes of mollusks came about initially as a method of increasing respiratory efficiency. The shell is thus an integral part of the molluscan respiratory system. The molluscan classes Gastropoda, Monoplacophora, Bivalvia, and Scaphopoda may have achieved their characteristic shell shapes as a consequence of building optimum configurations of pump gills.*

Because no representatives of the earliest mollusks still exist, there has been much speculation about what the "prototypical" mollusk may have looked like. This creature has even received a name: HAM, for Hypothetical Ancestral Mollusk. HAM is depicted in every invertebrate zoology and paleontology text. The creature is cephalized with a distinct head; there is the broad foot, internal viscera, and capping dorsal shell. Respiration is through the paired gills, located in a cavity at the back of the animal.

While the HAM model is but a theoretical construct, its overall morphology is at odds with the earliest mollusks found in the Cambrian, most belonging to lineages now extinct. Most of the earliest forms were minute, and in general their preservation is very poor, often simply as phosphate casts or molds in rock. Such preservation makes detailed anatomical reconstruction very difficult. Patient work by paleontologists Bruce Runnegar and John Pojeta has succeeded in classifying many of these earliest fossils, and the surprise is that the earliest do not resemble HAM in many ways. The largest group belongs to the monoplacophoran class, which may be ancestral to all subsequent mollusks. Long thought to have gone extinct at the end of the Permian, the discovery of living monoplacophorans in deep sea settings in the 1950s led to a much greater understanding of the life of the early mollusks. The living forms confirmed what muscle scars found on the interior of the earliest monoplacophorans fossils asserted— there was more than a single pair of gills. In fact, multiple pairs of muscles lined the entire length of the interior of the shell, leading to the conclusion that these early forms showed an evident segmentation or at least a repeat of the gill-blood vessel system. Since it is only the gills and supporting blood and filtering systems that show this repeated pattern, it can be surmised that, as in arthropods, this repeated pattern was an adaptation for the increased respiratory surface area of the gills. Today, a somewhat similar pattern of repetition, extending even to the shell, is found in the chitons, today commonly found on intertidal beaches. The point here is that the earliest mollusks seemed to have been designed with multiple sets of gills, necessary in a low-oxygen world of the Cambrian.

While the monoplacophorans flourished in earliest Cambrian strata and have survived to the present-day, other groups of mollusks soon evolved from the monoplacophorans. In them a reduction is seen in the number of paired gills, with ancestral gastropods and bivalves reducing the number of gills to a single pair and with primitive cephalopods maintaining perhaps two pairs, based on the anatomy of the modern-day *Nautilus*. So why was there a reduction of gills? As the reduction of the number of paired gills in mollusks occurred during the Cambrian, long before the models of Berner and others suggest that oxygen levels began rising, how can this observation be reconciled with the low-oxygen levels thought to have been present in marine environments during the earliest times of the Cambrian? Here it is proposed that the cephalopod shells evolved by the ancestral stocks provided a geometry that, like the monoplacophorans, allowed the formation of currents of water to be passed over the gills at a pressure that greatly increased the volume of water the gill surfaces would encounter in any given unit of time. In essence, the formation of these cephalopod shells, coupled with a ciliary system creating a unidirectional water flow across the gill surface, essentially increased the functional size of the gills. With this shell-ciliary system, a single pair of small gills became functionally equivalent to a much larger gill area of an organism with a "passive" gill system. Shells became part of the respiration organ in creating enclosed spaces or functional tubes for pressurized water to flow over the gills. This system also produced an effective method of passing used water out of the system, and the water flow also functioned to take out fecal and waste material at the same time.

Once in place, this new molluscan "shell-pump" form of respiration elaborated in several directions. The old passive method of multiple gills survived in the monoplacophorans and chitons, but neither group was ever very successful. But the pump gill forms showed a vastly different history, and here it is proposed that this huge success in terms of both species-level diversity and variety of form is directly related to the evolution and exploitation of this new kind of respiratory system. Three distinct elaborations on the pump gill respiratory system evolved in the Cambrian, and then each proceeded to undergo spectacular evolutionary flowerings producing numerous species. One group of mollusks rotated the body so that the gills faced forward into a water flow directed through the front of the shell's aperture and over the gills and exiting at the back of the aperture. These became the gastropods. A second group combined respiration and feeding by enlarging the gills even more with a pair of enclosing shells, using the bivalved shell morphology method of partitioning water flow across the gills in a fashion analogous to the brachiopods. This group became the bivalves. The third group left the gills in the back of the body but vastly increased the strength of the water flow over the gills by changing from a ciliary source of incurrent water propulsion to one caused by muscular pumping of the entire body, with each pump drawing water in across the gills and then using a return stroke forcibly ejecting the water back out of the shell. These became the cephalopods. In all three, shell form became a compromise between strength against predation and optimal shape for the channeling of interval water currents across the gill surface.

Let's look at the evolution of the cephalopods, which today include the familiar octopus and squid, in more detail. We know the *when* of their origin and even the *why*—the world was ready for the existence of highly mobile carnivores to prey on the abundant and slow-moving Cambrian arthropods—a stock of animals succeeded in evolving a body plan that could move fast despite the scarcity of oxygen.

While it has long been known that the cephalopod shell is a buoyancy organ, here is a new hypothesis: the nautiloid shell (and the wonderful shells of their future descendents, the ammonoids, creatures discussed in many future chapters) was first evolved as a respiratory organ in response to the low-oxygen content of the Cambrian oceans—and still functions like this today in what is the last remaining externally shelled genus.

All of this wonderful majesty of design came about because a late Cambrian mollusk discovered a new way of beating the problem of low oxygen: it evolved a new kind of molluscan respiratory pump, so that muscular action, instead of cilia, created a high-pressure volume of water that passed over the gills. After being mined for its oxygen and laden with carbon dioxide from respiration, all that water had to be expelled from the animal with as much force as it was brought in, if the same water was not to be recirculated. What better way than to jet it out through a tube! What a surprise it must have been when each jet of water jerked the entire animal backward. When predators came, an even more vigorous jet of water would jerk the earliest cephalopod (or was it still a monoplacophoran?) out of harm's way.

Natural selection then honed the system. But the shell was a heavy burden to bear, and thus a solution for neutral buoyancy was selected. When all was combined, in the late Cambrian, a new kind of animal was set loose. This was a hugely important episode in the history of life, as the cephalopods remained the dominant carnivores in the sea from that point until 65 million years ago when the ammonites, descendants of the cephalopods, were killed off by the Chicxulub asteroid strike, a time interval of about 450 million years. And as we now know, extinction of the ammonites was not the end of the cephalopods, for in large areas of the ocean, such as the pelagic or midwater regions, cephalopods remain the dominant carnivores, outcompeting fish in these dim regions of the sea. As we will see, cephalopods will be obvious and common players in the local marine ecosystems.

*Hypothesis 3.3: The molluscan class Cephalopoda, today comprised of squids, octopi, and the two externally shelled genera, Nautilus and Allonautilus (which still maintain the original cephalopods' body plan), produced its original distinctive shelled body plan as a way to build a highly efficient pump gill.*

The cephalopods are the pinnacle of invertebrate evolution in size and intelligence and rival the arthropods in complexity. Yet we are faced with a dilemma: the cephalopods have so radically departed from the characteristic body plan of head, creeping foot, and simple covering shell in both soft parts and hard parts that major explanations must be offered of how—and why—such radical changes occurred. When the major transition from a bottom-living ancestor to the swimming cephalopod, which first appeared in the late Cambrian (at about the same time as the Burgess Shale fauna, in fact), is discussed, the explanation is that the evolution of the buoyancy organ shell came first, followed by soft-part adaptations that would have perfected swimming—the jet propulsion system still used by squids today.

So here is the current concept. Long study of the living *Nautilus* has demonstrated that the iconic chambered shell does indeed function as a buoyancy organ, as first argued by Robert Hooke in the 1600s. The chambered portions are filled mainly with gas but with small volumes of water either present as free liquid at the bottom of each chamber or entrained in felt-like, water-containing membranes lining the interior of each septum and the siphuncle, the long calcareous (but porous) tube that pierces each chamber and contains a strand of living tissue connected to the back of the *Nautilus*'s soft parts. To make a new chamber, this posterior part of the body moves forward in the shell a short distance and the newly created space is filled with liquid, a blood filtrate. The *Nautilus* secretes a thin calcareous partition in front of this space and then progressively thickens (and thus strengthens) this new chamber. When sufficiently strong, an epithelium lining the siphuncle (which has been constructed of both hard and soft parts simultaneously with the secretion of the new chamber) begins to actively transport sodium and chloride ions out of the liquid filling the new chamber.

The chamber liquid in this way becomes progressively fresher compared to the blood in the siphuncular tube, thus producing an osmotic gradient. The chamber liquid, now relatively fresh compared to the blood, flows into the siphuncular tube and is passed by the blood vessels into the body of the *Nautilus* where it is excreted from the body. The removal of this liquid lowers the overall density of the animal with its heavy shell. It was long thought that, under pressure, the animal secreted the gas found in the chambers but this is not the case. Chamber pressure is never more than one atmosphere, regardless of the depth of the nautilus (and these animals have been caught at a depth of 600 meters, where the ambient pressure pushing against the shell is 60 atmospheres). This gas enters the shell by simple diffusion in response to the near vacuum conditions caused by liquid emptying (nature does indeed abhor a vacuum), but the gas plays no functional role. If liquid removal can balance the density increases caused by shell formation at the aperture and by growth of the soft parts (also denser than surrounding seawater), neutral buoyancy can be maintained.

So this is the major adaptation first seen in the Cambrian: A mollusk must have somehow built a calcareous wall at the back of its shell but left a strand of tissue within, which then had to evolve into an epithelial pump. So (the story goes) some poor late Cambrian monoplacophorans find themselves getting lighter and lighter and at some point float off the surface of the ocean bottom, off for an uncontrolled ride. Eventually, evolutionary forces shape the soft parts into tentacles and more importantly, shape a propulsion system created by the evolution of a jet of water forced through a tube-like funnel. Since the entire shell and animal is of neutral buoyancy, even a feeble squirt through the funnel would cause the animal to move off the bottom, presumably out of harm's way.

At any rate this is how the story has been understood. But in recent years many have become increasingly skeptical of this hypothesis, and eventually three lines of evidence have changed our views of the reasonableness of this scenario. First was the realization that the *Nautilus* does not use its buoyancy system for propulsion. For years, paleontologists thought that nautiluses, and by extension the many fossil ammonoids and *Nautilus* species, undertook nightly vertical migrations from deep to shallow water and did this by changes in buoyancy. Here a balloon analogy colored our view of things. A hot air balloon rises when new, heated air is vented into the balloon or if ballast is thrown out. In either case the density of the balloon is lessened, and the balloon rises in the sky. The balloon is brought back to Earth by venting the hot air. But much research showed that a powered dirigible is a

*Nautiloid cephalopods. This group evolved the consummate pump gill, and probably because of this became the largest of all invertebrate animals in the sea, with forms capable of thriving in water too low in oxygen for most other animals. The path of water into the animal, across its gills, and then out the funnel beneath the body allowed respiration and locomotion to be superbly combined.*

better model for the nautilus. In the grandiose Zeppelins of the early twentieth century, the balloon was kept at neutral and largely unchanging buoyancy in air and it rose or descended through its use of powered engines. So too with the nautilus. It was indeed discovered that the nautilus undertakes nightly vertical migration, but changes in buoyancy through trim of the shell are not involved. Instead, the strong swimming action of water jetted through a tube beneath the shell powers the animal upward and also pushes it back down to depth.

A second line of discovery dealt with the jet power itself. It is enormously strong, and as scientists made further measurements of the system, it was discovered that a large volume of water was constantly being pumped through the front of the shell, even when the animal is at rest and motionless. There is a simple reason for this—the propulsion system is an offshoot of the respiration system. All of this water that is eventually destined to jet through the locomotion tube, called the hyponome in the *Nautilus* and *Allonautilus,* first passes over two pairs of large and complex gills in the back of the mantle cavity. The design allows a dual function—respiration and locomotion—from the same energy expended to draw water in and then force it out.

Finally, in the 1990s it became clear that the *Nautilus, Allonautilus,* and other cephalopods as well are very efficient at respiration, even in seawater that has oxygen levels lower than that of normal seawater. New research now shows that the *Nautilus* commonly visits low-oxygen environments. A nautilus can be removed from water for a half-hour with little ill effect. Squid have recently been observed to enter low-oxygen water masses in the Gulf of Mexico with impunity, places where fish cannot go. The system that allows movement has made these animals champion respirers. These three challenges to the earlier view of *Nautilus*'s evolution being driven by the need for neutral buoyancy has led me to a new model for *Nautilus*'s Cambrian evolution.

Cephalopods thus came about following the evolution of a marvelous and efficient gill, one that allows them even today to visit anoxic water masses and still harvest whatever oxygen that can be found there. This was the secret of their success, and later in time, when oxygen levels rose, they became even more efficient. They always had a better respiratory system than their prey and competitors, and the doubling up of the respiratory system with locomotion sealed their success. Eventually most lost their shells, but the respiratory system of the cephalopods remains supreme and will ensure their existence far into the future.

## WHY CHORDATES EVOLVED AS THEY DID

Having looked at the respiratory structures of the most populous Cambrian animals, it is time to take a look at a small and insignificant group of fossils found at Chengjiang and later at the Burgess Shale—small, fish-shaped animals that ultimately gave rise to us. The most famous of these was named *Pikaia*, an animal featured at length in Steven Jay Gould's book about the Burgess Shale animals called *Wonderful Life*. The origin of our own phylum is, of course, of intense interest and has been the subject of acrimonious debate for decades, caused in no small way by the dearth of fossils from the time of presumed chordate, or vertebrate, ancestry. Data pertaining to the origins of our phylum come from comparative anatomy and development of living representatives of various phyla seemingly related to us chordates, from DNA studies, and more recently from the interesting new fossils found in Early Cambrian strata in Chengjiang, China.

Compared to other phyla, ours is very different in several anatomical characteristics. Most animal phyla that show bilateral symmetry have a nerve chord running the length of the body, but this nerve cord lies beneath the gut. In chordates, it is dorsal to the gut. Getting to this reversal of anatomy required some major evolutionary rearranging. One of the oldest and most elegant hypotheses came from the evolutionist W. Garstang, who noted the similarity between the presumed anatomies of ancestral vertebrates (and the fish-like but boneless Lancelet, *Amphioxus*, which is often used as a model of what the first chordates might have looked like) and the larval stage of the common tunicate, commonly called a sea squirt. While tunicates are sessile filter feeders and look nothing at all like any sort of living chordate, their larva strongly resemble small fish. It was thus theorized by Garstang, followed later by a slew of anatomists, that true vertebrate chordates arose from the larva by a process called paedomorphism—where evolution causes larval characteristics to appear in adults. Later DNA studies have supported this view. We now have a pretty good idea about the "tree" of evolution leading to us chordates: our nearest nonchordate ancestor appears to be the phylum Urochordata, the phylum containing the sea squirts, as so long ago suggested by Garstang.

Is there a connection between this phylogeny, or proposed evolutionary pathway, and levels of oxygen in the latest pre-Cambrian, when this split of the tunicate group into tunicates and vertebrates probably occurred? There has never been any published suggestion that oxygen levels had anything to do with the evolution of the phylum Urochordata. Let's change that now:

*Hypothesis 3.4: The phylum Urochordata—ancestor of the chordate—evolved in response to low oxygen by producing a body plan with a highly powerful and efficient pump gill, which became co-opted for feeding as well. It was this body plan that led to the larva, which became a template for the chordate body plan. The chordates thus came into being because of the body plan of their immediate ancestors.*

To support this hypothesis we need to look in some detail at the anatomy of the tunicates, present-day representatives of the Urochordata. A tunicate looks like anything but a vertebrate at first glance. Tunicates are sessile filter feeders with two prominent "chimneys" on their topside. Water is sucked into the first tube, forced across a large gill/feeding filter, and then forcibly jetted out of the second tube. The pump itself is powered by muscular action of the outer body wall. Some tunicates move several thousand times their body volume of water through this system daily. According to new, unpublished calculations, this makes tunicates among the champion pump respirers in all the animal kingdom.

Almost the entire inside of the tunicate's sack-like body is filled with an enormous gill structure. It is highly subdivided, creating a large and intricate surface area for respiration. Blood moves through the interior of these gills, allowing a highly efficient respiratory exchange and is moved through a circulatory system by a heart. Tunicates have no blood pigment, and their inactive life style requires little oxygen. At first glance, their gill system seems to be a case of overkill—there is far more potential than need. In our highly oxygenated oceans it has been estimated that tunicates extract only 10 percent of the available oxygen, which is all they need. If ever there was a design that can handle very low oxygen, this is it.

The respiratory organ inside the adult tunicate has a series of slits within it, and during development these same slits are formed in the front of the tadpole-like larval stage. The larva settles on the bottom and metamorphoses into the sessile adult. The gill slit system of the larva is essentially a perfect preadaptation for the gill system found in fish because the gill structure in the adult tunicates is of a morphology that is fortuitously and easily changed through evolution into the familiar fish gill. The fossil record from Chengjiang contains fully evolved and recognizable tunicates, indicating that tunicates were certainly there well back in the Cambrian.

Although some scientists speculate that the pump may have evolved to capture food when the body was small enough to respire more passively, rather than for respiration, our perspective emphasizes the significance of low-oxygen levels in driving evolutionary responses to respiratory requirements. Here is the new reasoning behind Hypothesis 3.3. As we have seen, the very low oxygen of the late Precambrian stimulated many morphological ways to get around the obstacle of low oxygen. What little oxygen that was present had to be extracted from the water, and to do this, animals developed a host of morphological solutions. The solution of the tunicate was one of the most elegant and effective. It was also of a design that could be modified for use in a bilaterally symmetrical, swimming creature—the ancestor of our group. Thus, we chordates owe our existence to a solution to the low-oxygen problem confronted by the founding animals involved in the Cambrian Explosion.

## THE END OF THE CAMBRIAN

The Cambrian Explosion was a time of rapid diversification. But near the end of the period, that push toward ever-greater numbers of different types of animals stopped and diversity began to drop. A mass extinction, perhaps the first ever encountered by animals, descended on life; it's quite clear there was a series of low-oxygen events in the sea. The trilobites were seriously affected: whole families disappeared, and the survivors and newly evolved forms of the first part of the succeeding Ordovician had a different morphological look. Also killed off were many of the exotic arthropods that had lived during the period, in-

*Reconstruction of* Pikaia, *one of the earliest chordates.*

cluding *Anomalocaris*. It's a shame that there is not a Burgess Shale–like deposit in the earliest Ordovician rocks so that we could really take a census of how many soft-part animals known from the Cambrian-aged Burgess Shale made it through this biotic crisis. It may be that the end-Cambrian event was every bit as destructive as any of the Big Five—but that it has remained underestimated because fewer animals at the time had skeletons and were thus less likely to have been preserved. Whatever its destructiveness, one thing seems clear: the extinction appears to coincide with a rapid drop in oxygen, and this drop may be related to fundamental changes in the carbon cycle. As noted in Chapter 2, sudden drops in oxygen were mass extinction instigators. But the extinction was also an instigator of future diversity, if the relationship noted in Chapter 2 about low oxygen stimulating new species formation is correct.

It would be fascinating to compare the respiratory structures in those animals going extinct at the end of the Cambrian to those that survived. My suspicion is that the survivors were more efficient respirers than those that died out. This is research for the future, however.

## ORDOVICIAN REBOUND

The oxygen drop at the end of the Cambrian was short lived. There was a rebound in atmospheric oxygen levels, and with the rebound came a higher diversity of life in the succeeding Ordovician period, coming, perhaps, from a variety of new body plans evolved in the crisis of low oxygen at the end of the Cambrian. This Ordovician expansion of life is the subject of the next chapter.

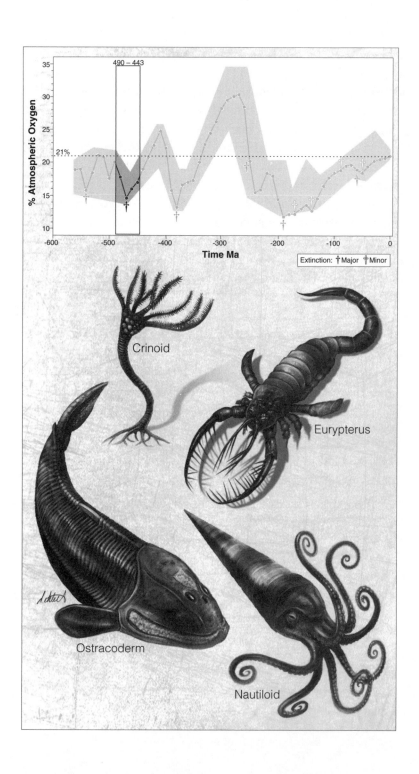

Crinoid

Eurypterus

Ostracoderm

Nautiloid

# 4

## THE ORDOVICIAN:
### CAMBRIAN EXPLOSION PART II

The Ordovician Period followed the Cambrian and it is recognized as a separate time interval because of differences in the kinds of animals found in each. As we saw in Chapter 3, the Cambrian came to an end because of a mass extinction. After any mass extinction there is usually a time of rapid evolution to fill empty ecological niches, and so it was with the Ordovician. Many kinds of animals were present in the Ordovician that had not yet evolved in the Cambrian, and many of these appeared soon after the end of the Cambrian mass extinction. The result was an assemblage of animals that were markedly different than most of the Cambrian faunas. Trilobites were still there, but they were swamped out by what might be called a "shelly fauna" of invertebrates, dominated by the bivalved brachiopods, with lots of corals, bryozoans, echinoderms, and mollusks thrown in. It is here too that the first coral reefs appeared as well as the first skeletonized fishes. Let's go back and look at how the ancient Ordovician world might have been.

### TRIP 2

We have dusted off our time machine for a look at the early Ordovician. As in the previous chapter's voyage, this machine is part airplane and part submarine—and entirely imaginary. What we might see, however, comes from two centuries of paleontological work.

We sweep down over the land surface and notice immediately that it is greener. There are still no trees or even commonly rooted plants, but mosses have spread and among them are primitive vascular plants. There is also the occasional movement of animals on the land surface, not vertebrates, but a diversity of scorpion and centipede-like arthropods. There is not yet a high diversity of them, but there are enough to convince us that an invasion has started, a beachhead has been established, and more troops are on the way. As we pass over the landscape, much is familiar: mountains and plains, great glaciers in the high latitudes, sandy beaches with sand dunes and ripples, the spray of the sea on the land. True, the lack of trees and bushes, flowers and grass, birds and even flying insects is curious enough, and there is much more bare rock than we are used to outside the desert, high mountain, or glacial regions of our own world. But there is something else about this landscape that nags, something also seen on our previous look at the land in the Cambrian world 50 million years previously. As during the Cambrian, the rivers are still braided, not meandering. There are no riverbanks, just a vast wide complex of shallow rushing water on its journey from the distant snow-capped mountains to the sea.

We return to the seashore and power downward into the sea, expecting changes and not being disappointed in this. At first, as we settle onto a shallow, warm bottom, things look superficially the same because the most common bottom dwellers are sponges. In the clear blue tropical water, we first think we are in the shallows off Jamaica or the Florida Keys—there are sponges everywhere, in myriad colors and shapes. Many are glass sponges, but the more familiar demosponges are there too. Amid them are untold numbers of bivalved creatures, looking a bit like clams of our world. But closer inspection shows them to be brachiopods, forms with articulated shells that are more advanced than the small, inarticulate brachiopods of the Cambrian world. They coat the bottom, and amid them is an animal not seen in the Cambrian—the colonial bryozoan, which builds calcareous skeletons, each being a tiny miniature replicate of the solitary brachiopods. The brachiopods look like clams from our world, but this is but a trick of evolution, where successful body designs can be seen in un-related stocks. Both bryozoans and brachiopods show a large feathery device used for feeding—the lophophore. But while acknowledged as a feeding organ, it is evident that it is used as a respiratory organ as well. Both the solitary brachiopods and the colonial bryozoans are seen to be actively moving water across these lophophores in impressive fashion. For the brachiopods, water is sucked into the shell and then blown out after crossing the lophophore. The colonial bryozoans pro-duce more complex but equally effective water currents by position-ing the tiny zooids, thousands of which make up medium-sized colonies, in such a position as to produce an area on the colony where water enters the forest of tiny lophophores and then different areas where this water is blown away from the colony. For all such animals, once the water has been mined of its twin treasures, food in the form of microscopic plankton and bacteria and life-ensuring oxygen, it must be sent away so that it is not recirculated.

We continue onward, and a great stony city comes into view. It is a reef, and it looks, at least at first glance, surprisingly familiar. We are 475 million years in the past, hovering over a reef that would make a Club Med proud. The size and shape of this reef are something that we did not see in the Cambrian. There were reefs back then, but they were made up of the remarkable archeocyathids, sponges that build cylin-drical and vase-shaped skeletons and are very much smaller in size. This one is immense, a great stone edifice of life. Here we see three dominant kinds of reef builders—colonial corals, the tabulates; large calcareous sponges called stromotoporoids; and solitary corals with horn-shaped skeletons, the rugose corals. It is remarkable—when we look at the whole, the shape of the reef through squinted eyes is so familiar. The same coral-rich walls in the fore-reef areas, the same reef flats, the back-reef lagoon areas filled with animals—colonial and soli-tary. But then we look at what is making this structure, and it is all different. The reef community here is made up of an entirely different assemblage of builders and binders than what will be the bricks and mortar of the reefs in our time, yet the overall reefs look much the same, just different bricks. Is there a respiratory function to forming these big colonies? Possibly. In our own world the very shape of the reefs themselves (not just the coral that makes them) maximizes water movement over the millions of small animals living there—and in so doing brings in more oxygenated water than would otherwise be present. Back in the Ordovician the same may apply, for the reef shapes (but not the corals making them) are very familiar indeed.

But is it really the same? It takes time to record all the impressions, to make the connections of similarity and difference. There is oddness here—and then it hits home: no fish.

The swarming schools of reef fish, the amazing color and diversity in our world's reefs, are nowhere to be seen back in these Ordovician-aged seas. There are a few swimmers, plenty of trilobites around, and boundless numbers of other arthropods busily going about their business of life. Many bear a cornucopia of upward-pointing spines, bearing witness to the dangers in this world. These are defensive adaptations, all, and we immediately start searching for the familiar top carnivore of the older world, the anomolocarids that so effectively ruled the top of the Cambrian ecological pyramid (producers eaten by grazers, in turn eaten by carnivores) for so long. But they are nowhere to be seen and in fact are by now long extinct. Two things happened. There is now more oxygen in the Ordovician, and a new scourge of a predator has appeared in the seas. We look again into the blue waters, and among the swimming arthropods we readily see an entirely different kind of animal up in the water column. It is a body plan that was not present at all in the early and mid-Cambrian, one that first appeared only at the very end of the Cambrian in fact, at the same time that the mass extinction of trilobites and other Cambrian arthropods was under way. We look more closely at these strange swimming predators and see an animal type that is virtually unknown in our world. These are shelled creatures, but unlike the body armor of the arthropods, which is composed of many segments, here the conelike shells are the most striking feature, some straight, some gently curved, some entirely coiled. Then we see something very similar to the chambered nautilus of our world and realize that we are seeing an amazing variety of chambered cephalopods. In only a few minutes of watching their activities, a new conclusion is reached. These nautiloids have toppled the arthropods as the top carnivores.

While some are but a few inches long, there are also veritable battleships, long straight shells passing backward through the water. Some are immense. Six footers are common, but one is easily double that size, and from the large round opening of its conical tubular shell a great head, with saucer eyes, watches, searches, looks. A mass of tentacles surrounds the head, and we can see that propulsion is from great volumes of water being jetted from a funnel-like fleshy tube situated beneath the mass of tentacles. With shell and flesh, the entire animal must weigh several hundred pounds in air, but in this sea it is weightless. Although invisible, we know the inside of the massive shell is composed of numerous chambers, filled with air, but with a thin tube containing a strand of blood vessels passing through the middle of each chamber. The nautiloids have discovered the secret of achieving neutral buoyancy. But they have done far more than this. They have climbed to the top of the trophic pyramid, and it may be that it was the best respiratory system on the planet that got them there.

We move from the sea and on a whim take one more dive—into the brackish water of a large estuary connected to one of the extensive braided river systems. Here we see in large numbers some familiar shapes. There are fish in this fresh water. But they are fish very foreign to us. They are called ostracoderms, and they have small sucking mouths without jaws. Most have extensive dermal body armor in the form of massive scales and larger bony plates. Their tail is "reversed heterocercal," like an upside-down shark's tail. And no wonder, most of these early fish are swimming at the surface, thanks to the action of these tails whose design naturally forces the body upward, and they are eating the prolific algae and vegetable scum at the lagoon's surface. A few other types are seen as well, hugging the bottom, and these are even more heavily armored than their surface-living brethren.

We look more closely at some of these primitive fish for evidence of a respiratory structure. One of these fish, *Arandaspis*, shows 15 plates covering a primitive gill system called gill pouches. The gills themselves are relatively large compared to the size of the fish, and swimming seems part of the respiratory cycle. Forward motion passes water through the gills, making this the functional equivalent of a pump gill.

*Reconstruction of a fish typical of the Ordovician period, the armored genus* Hemicyclopsis. *These types of fish, known as ostracoderms, did not have jaws. Their closest living relatives are the lampreys and hagfish.*

## OXYGEN, CARBON DIOXIDE, AND ORDOVICIAN EVENTS

The Ordovician Period can be regarded as the second half of the two-part initiation of animal diversity on Earth, with the first being the Cambrian Explosion. There is a good reason for this. Like the Cambrian, it was a time when new species as well as new kinds of body plans appeared at a faster rate than was characteristic of more recent times. This high rate was in response to filling up the world with animals for the first time.

While the Cambrian ended with a series of minor extinction events affecting mainly trilobites, this dip in diversity was short lived and was succeeded by a huge increase in the number of animal species in the sea, especially among calcareous shell-forming organisms. As we saw on our trip, large coral and sponge reefs appear, as does the first extensive plankton made up of animals—in this case, floating colonies of tiny filter feeding animals called graptolites, which are now totally extinct. Was there a new trigger to this diversification or was it but a continuation of the Cambrian Explosion? Here again we can invoke the record of oxygen levels to understand this rise in diversity of animals, as well as the first appearance of communities that were largely composed of calcareous skeleton-forming organisms. Skeletal building (as well as every aspect of animal physiology, as we saw in Chapter 1) works better in air or water that is rich in oxygen and nutrients.

But the Ordovician was a time when oxygen levels were still lower than today, and in the middle part of the period they were markedly so. The extensive appearance of many animals with calcareous shells, some, like the corals, massive in size, indicates that there was significant oxygen in the oceans during some of the period. But there was also a major mass extinction, one of the Big Five. This mass extinction occurred either simultaneous with or soon after a drop in oxygen levels to their lowest levels of the period. While gains in diversity made in the Cambrian Explosion were consolidated over part of the Ordovician, this major drop in diversity may have been related to a profound oxygen drop.

How important was respiration in shaping the kinds of animals of the Ordovician? Very important, in my opinion. It takes an ever-better respiratory system to deal with ever-larger bodies. The Cambrian was a time of mainly small animals, most less than an inch long. The Ordovician was a time of much larger animals. For example, in the Ordovician the brachiopods, all mollusks (most notably the giant nautiloid cephalopods), echinoderms, the chordates (our group), and many arthropods were noticeably different if each group as a whole is compared to Cambrian-aged members. In each case this bigness came about through changes and improvements in respiratory systems.

Let's now move forward in time, looking at events of the Silurian and Devonian, when a short-lived spike in oxygen to the highest levels ever attained on Earth up to that time caused major events in the history of life. Until now, we have explored how low oxygen shaped the biological world. The next chapter will explore the reverse, looking at the consequences of rising oxygen levels at the time when animals and plants colonized the terrestrial realm in earnest.

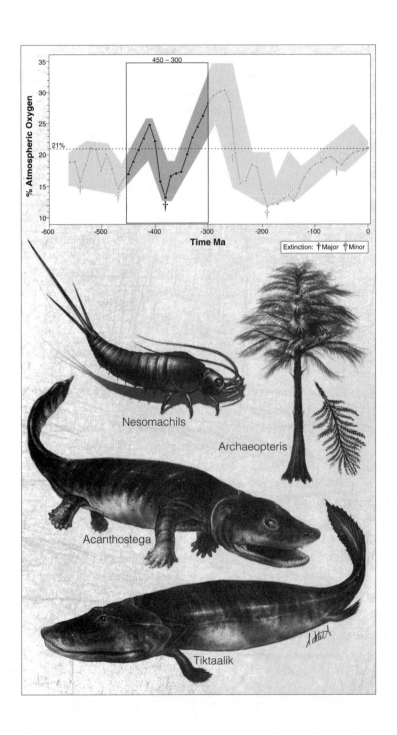

450 – 300

35
30
25
21%
20
15
10

% Atmospheric Oxygen

-600   -500   -400   -300   -200   -100   0

**Time Ma**

Extinction: † Major  ✝ Minor

Nesomachils

Archaeopteris

Acanthostega

Tiktaalik

# 5

## THE SILURIAN-DEVONIAN:
## HOW AN OXYGEN SPIKE ALLOWED THE
## FIRST CONQUEST OF LAND

Chapter 4 looked at the world of the Ordovician, a time interval beginning with a wave of extinctions, which ended the Cambrian period some 490 million years ago, and ending with less violence some 450 million years ago. This chapter looks at the next two periods: Silurian (443 million to 416 million years ago) and Devonian (416 million to 359 million years ago). During this time, great reefs grew in the seas; fish began to vie with the cephalopods for dominion in the oceans, while untold numbers of brachiopods covered the warm sea bottoms. Yet for all of the changes in the ocean, it was on land that the greatest transformation took place. For the first time animals began to trek inward from the seashore, breathing through new kinds of lungs as they emerged from their beachheads. They did so in search of new food sources, for a revolution was taking place: the land was greening, and plants, for the first time, rose up from the soil to point green leaves at an enlarging sun. Why did this invasion of the land take place? The old view was that it was about time, nothing more. Here, in this chapter, let's join those ancient plants and early land animals in breaking new ground. From the middle of the Ordovician Period until the end of the ensuing Silurian Period—a time of 60 million years—atmospheric oxygen uninterruptedly rose 10 percentage points, from a gasping 15 percent to the heady richness of 25 percent of the air—far more than we have today. The land changed from a place of "thin air" to a world where oxygen was almost a free commodity—which would have been a wonderful and necessary thing for an early insect with a still nearly worthless, newly evolved lung or an early land plant trying to coax oxygen into one of its clumsy, inefficient, newly evolved roots, for while plants above the ground need little oxygen for their leaves, their roots, deep in soil that may have too little oxygen for growth, are another matter. Was this coincidence or cause and effect? In this chapter let's look at how this great rise in atmospheric air paved the way for the first real colonization of land.

### TRIP 3

Let's travel back in time to the last time interval of the Silurian Period, a time some 420 million years ago, to get a snapshot of the middle of this long Silurian-Devonian interval. It has been some 50 million years since our last look at Earth, and the changes that have taken place are nothing short of remarkable.

The land is green, and everywhere we see the stems of low plants not only at water's edge of swamps and lakes but also in uplands. There are true vascular plants, still low in size but now common and diverse: an enormous variety of forms are seen. And they are not alone on the land. Crawling among the plants is a small diversity of insects, also very small in size. They look like today's springtails and the odious silverfish, and none have wings. Among them are even smaller mites. There are no land-dwelling vertebrates of any kind—yet.

We move to a small lake and find it filled with fish. For the first time we see fish with jaws amid the same jawless ostracoderms that were present in the earlier Ordovician. The jawed fish are heavily armored and upon closer examination seem very foreign looking. They are the placoderms, and some have gotten quite large and are destined to become larger yet in the upcoming Devonian Period. They use their jaws to good use, making meals of the algae-eating and jawless ostracoderms. There are a few minnow-like fish that seem to look like the bony fish of our time and perhaps some fish that may become sharks, but these are rare. The bony armor seems like a ludicrous adaptation for animals depending on mobility and maneuverability, but soon the reason for this armor becomes apparent. A great shadow looms from the turbid water, and a hideous nightmare swims into view. It is clearly some kind of arthropod, coated with the typical arthropodan external skeleton, but its size is staggering. It looks like a giant underwater scorpion and is indeed related to the scorpion clan. Many others swim into view, undulating segmented bodies like human swimmers doing the butterfly, but these animals are propelled through the water by a large horizontal tail. The largest of these creatures, called eurypterids, is 10 feet long. They are the largest animals yet evolved on Earth and one of the most fearsome of any that will ever inhabit the sea.

The eurypterids sweep into the slowly moving fish, and their means of attaining food is apparent. Each has a long pair of wicked claws emerging from the head region, and many fish become meals. How could any animal be so large at this time? Up until now animals had been far smaller. We know that the amount of oxygen available to an animal is often a determinant of its final size. A quick measure of atmospheric oxygen tells the story: it is 24 percent compared to our present-day 21 percent. A realization dawns: no giant arthropods like this exist in our time, nor could they, in all probability. These giants are enabled by high oxygen. But this time of oxygen abundance will not last, and by the end of the succeeding Devonian Period oxygen levels will crash to far lower values than today. With this crash, the race of arthropod giants and many, many more animal races will be killed off in one of the greatest of all mass extinctions affecting animals (and plants) on Earth—the Devonian mass extinction of 368 million years ago. But that is still far in the future, over 50 million years in the future. For now the high oxygen of the Silurian has allowed the rise of the first insects on land, the rise of plants on land, and the first age of giants here in fresh waters. We wonder what awaits us in the sea.

The changes there are no less dramatic. The shelly marine fauna of sessile marine invertebrates is still in place, but again we remark on the size of things. On a shallow sea bottom we find a pavement of shelled brachiopods, looking much like the clams that also occur in some numbers around them. But these brachiopods are the largest that we have ever seen and, like the eurypterids, the largest of their kind that

Earth will ever see. They are pentamerid brachiopods, and, like the eurypterids, their time is short, for the fall of oxygen soon to come will spell their end. The high atmospheric oxygen has translated in this shallow sea bottom to a very high level of dissolved oxygen, and the brachiopod pump gill makes these animals well fit for living in this high-oxygen world and allows this gigantism.

We see fish here too, many resembling those of the freshwater lake we visited, but the fish are both outnumbered and dwarfed by an incredible diversity of nautiloid cephalopods. The cephalopods vary in size from tiny to gigantic, some larger even than the eurypterids, and there is a cornucopia of shapes as well. For the cephalopods, too, the high oxygen has translated into enormous size—and this makes good evolutionary sense. (The nineteenth-century paleontologist Edward Cope realized that most lineages of animals become larger through time—and he recognized the reason. Large size lends protection against predation. This generalization now bears his name: Cope's Rule.) The long, straight cones typical of the first nautiloids are still present, and the biggest of the nautiloids are all straight. But among the cephalopods are all manner of shapes from arcuate, or snail-shaped, to tightly coiled forms not too dissimilar from the living *Nautilus* of today.

We look for the prey of the nautiloids and see a variety of trilobites. When threatened, the trilobites coil up, but the most striking thing about their appearance is that most sport far fewer thoracic segments than were present in their Cambrian and even Ordovician ancestors, and many lineages have become larger as well. With this reduction of segments there has been a vast reduction of gill surface area, but once again, the high-oxygen content of these marine waters has allowed the trilobites this luxury. But they have made a bad bet in doing this, and in many respects this time of higher oxygen is their last hurrah. The great Devonian mass extinction will kill off most, and while a few will survive the catastrophe, they will from that point on, until their final extinction in another mass extinction (the most terrible of all, at the end of the Permian), remain rare and minor elements of the marine world.

## TERRESTRIALIZATION

While numerous innovations took place in the sea during the Silurian and Devonian periods, it is on land that the most important events—at least to us land-living animals—took place. The conquest of land, first by plants and then by a succession of animal phyla, began in the Silurian and Devonian. Why did the conquest of land occur then? The standard view of the history of life is that these events took place because animals had finally evolved to a point where land conquest was possible. In other words, the evolutionary advances in arthropods, mollusks, annelids, and eventually vertebrates—the major animal phyla involved in the conquest of land—had finally and coincidentally arrived at levels of organization that allowed them to climb out from water and conquer the land. An alternative view is that the first conquest of land took place as soon as atmospheric oxygen rose to levels allowing land animal life. Let's first look at what was required of both plants and animals to allow terrestrialization, the adaptations needed to permit life on land. Let's begin with plants, for without a food source on land, no animals would have made the effort to gain a terrestrial foothold.

By 600 million years ago, plant evolution had resulted in the diversification of many lineages of multicellular plants, some of which are familiar to us still—the green, brown, and red algas, which are members of any seashore in our world. But these were plants that had evolved in seawater. The needs of life—carbon dioxide and nutrients—were easily and readily available to them in the surrounding seawater. Reproduction was also mediated by the liquid environment. The move to land required substantial evolutionary change in the areas of carbon dioxide acquisition, nutrient acquisition, body support, and reproduction. Each required extensive modification to the existing body plans of the fully aquatic plant taxa.

It was the green algal group, the *Charophyceae*, which ultimately gave rise to photosynthetic land plants. Many obstacles had to be overcome. First was the problem of desiccation. Green algae washing ashore from their underwater habitat quickly degenerate in air, for they have no protective coating. But these green algae produce reproductive zygotes that have a resistant cuticle, and this same cuticle may have been used to coat the entire plant in the move onto land. But the evolution of the cuticle, which protected the liquid-filled plant cells inside, created a new problem: it cut off ready access to carbon dioxide. In the ocean, carbon in dissolved carbon dioxide was simply absorbed across the cell wall. So to accomplish this in the newly evolved land plant, many small holes, called stomata, evolved as tiny portals for the entry of gaseous carbon dioxide.

The plant body must be anchored in place, and early land plants were probably anchored by fungal symbionts. Additionally, the symbiotic relationship between plants and fungi would provide for a means through which water could be recovered from the soil.

Moving onto land also created the problem of support. Plants need large surface areas facing sunlight so that their chloroplast receive enough energy through light to run the photosynthetic reactions necessary for plant life. One solution is to simply lie flat on the ground, and the very first land plants probably did this. This kind of solution is still used by mosses, which grow as flat carpets lying over soil. A visit to the Ordovician land probably would have been a visit to a moss world, where the world's tallest "tree" was all of a quarter-inch tall. But this is a very limiting solution. Growing upright enables the acquisition of much more light, especially in an ecosystem where there is competition between numerous low-growing plants, and harder material was incorporated by early plants to allow first stems and finally tree trunks. Concomitant with the evolution of stems would have been the evolution of a transport system from the newly evolved roots up to the newly evolved leaves. Finally, reproductive bodies that could withstand periods of desiccation evolved, enabling reproduction in the terrestrial environment. With these innovations the colonization of land by plants was ensured, and with the formation of vast new amounts of organic carbon on land for the first time, animals were quick to follow. New resources spur new evolution.

As with plants, a major problem facing any would-be terrestrial animal colonist is water loss. All living cells require liquid within them, and living in water does not create any sort of desiccation problem. But living on land requires a tough coat to hold water in. The problem is that solutions that allow a reduction in surface desiccation are antagonistic to the needs of a respiratory membrane. So—to build an external coating that resists desiccation and then suffocate? Or build a surface respiratory structure that allows the diffusion of oxygen into the body but risk desiccation through this same structure? This dilemma had to be overcome by any land conqueror, and it was apparently so difficult that only a very small number of animal phyla ever accomplished the move from water to land. Some of the largest and most important of current marine phyla certainly never made it: there are no terrestrial sponges, cnidarians, brachiopods, bryozoans, or echinoderms, among many others, for instance.

Identification of the first land animals has relied on a fossil record that is notoriously inaccurate when it comes to small terrestrial arthropods. The oldest fossils of land animals all appear to be small spiders, scorpions, or very primitive insects dating back to between 420 million and 410 million years ago—right about the transition from the Silurian to the Devonian. All of these animals, however, have very weakly calcified exoskeletons and are rarely preserved. By the late Silurian or early Devonian, however, the rise of land plants also brought ashore the vanguards of the animal invasion, and it is clear that multiple lines of arthropods independently evolved respiratory systems capable of dealing with air. The respiratory systems in today's scorpions and spiders provide a key to understanding the transition of arthropods from marine animals to successful terrestrial animals. Of all body structures required to make this crucial jump, none was more important than respiratory structures. It also seems apparent that the earliest lungs used by the pioneering arthropods would have been transitional structures nowhere near as efficient as the respiratory structures in later species. But in a very high-oxygen atmosphere, air can diffuse across the body wall of very small land animals—and the first land animals all seemed to be small—and can enter the body through even the primitive lung structures.

Of the phyla that made it onto land—the arthropods, mollusks, annelids, and chordates (along with some very small animals such as nematodes)—the arthropods were preevolved to succeed, for their all-encompassing skeletal box was already fashioned to provide protection from desiccation. But they still had to overcome the problem of respiration. As we have seen, the outer skeleton of arthropods required the evolution of extensive and large gills on most segments to ensure survival in the low-oxygen Cambrian world (where most arthropod higher taxa are first seen in the fossil record). But external gills will not work in air. The solution among the first terrestrial arthropods, spiders and scorpions, was to produce a new kind of respiratory structure called a book lung, named after the resemblance of the inner parts of this lung to the pages of a book. A series of flat plates within the body have blood flowing between the leaves. Air enters the book lungs through a set of openings in the carapace. The book lung is a passive lung in that there is no current of air "inhaled." And because they have a passive lung, animals with book lungs depend on some minimum oxygen content. Surprisingly, no experiments have yet been done to see how low they can go.

Some very small spiders are blown by winds at high altitudes and have been dubbed "aerial plankton." This would seemingly argue that the book lung system in spiders is capable of extracting sufficient oxygen in low-oxygen environments. But these spiders are invariably very small in size, so small that an appreciable fraction of their respiratory needs may be satisfied by passive diffusion across the body. Larger-bodied spiders are dependent on the book lungs.

Book lungs may be more efficient at garnering oxygen than the insect respiratory system, which is composed of tube-like trachea. Like the respiratory systems of spiders and scorpions, the insect system is passive in that there is little or no pumping, although recent studies on insects suggest that some slight pumping may indeed be occurring but at very low pressures. The book lung system of the arachnids has a much higher surface area than does the insect system and thus should work at lower atmospheric oxygen concentrations.

Understanding the *when* of this first colonization of land is hampered by the small size and poorly fossilizable nature of the earliest scorpions and spiders. Present-day scorpions are more mineralized than spiders, and not surprisingly, have a better fossil record. The earliest evidence of animal fragments on land is from late Silurian rocks in Wales, about 420 million years in age, near the end of the Silurian Period—a time when oxygen had already reached very high levels, the highest that until then had ever been present on Earth. These early fossils are rare and of low diversity, but identifications have been made: most of the material seems to have come from fossil millipedes.

A far richer assemblage is known from the famous Rhynie Chert of Scotland, which has been dated at 410 million years in age. The Rhynie Chert deposit has furnished fossils of very early plants, and from these cherts, small fossil arthropods are known as well. The arthropods are mites and springtails, which both eat plant debris and refuse. Mites are related to spiders and are very small in size. Springtails, however, are insects, presumably the most ancient of this largest of animal groups on Earth today. While it might be expected that once they evolved, insects diversified into common elements of the early Devonian fauna, just as they make up the most common element of terrestrial animal life in our time, actually the opposite appears to be true. According to paleoentomologists, insects remained rare and marginal members of the land fauna until nearly the end of the Mississippian Period, some 330 million years ago.

Insect flight also occurred well after the first appearance of the group, with undoubted flying insects occurring commonly in the record some 330 million years ago. Soon after the first development of insect flight, the insects undertook a fantastic evolutionary surge of new species, mainly flying forms. This was a classic "adaptive radiation," where a new morphological breakthrough allows colonization of new ecological niches. But that radiation also took place at the oxygen high 330 million years ago near the end of the Mississippian Period— when oxygen levels had reached modern-day levels and in fact were on their way up to record levels that climaxed in the late Pennsylvanian Period of some 310 million years ago—and was surely in no small way aided and abetted by the high levels of atmospheric oxygen.

So were insects the first animals on land? Surely not, according to those who study early terrestrial animal life. That accolade may go to scorpions. In mid-Silurian time, some 430 million years ago, a lineage of proto-scorpions with water gills crawled out of the water and moved about on land, perhaps scavenging on dead animals washed up on beaches. Their gill regions remained wet though, and the very high surface area of these gills may have allowed air respiration of sorts. So

*Reconstruction of the early terrestrial arthropod* Archaeognatha, *a flightless true insect.*

we see a timetable: scorpions out first at 430 million years ago but of a kind that may have been still tied to water for reproduction and perhaps even respiration, followed by millipedes at 420 million years ago, and insects at 410. But common insects did not appear until 330 million years ago. How does this history relate to the atmospheric oxygen curve?

The Berner curve for this time interval (shown at the start of this chapter) indicates that the end of the Silurian was a time when oxygen had already reached very high levels—the highest that until then had ever been evolved on Earth—with a high-oxygen peak at about 410 million years ago, followed by a rapid fall, with a rise again from very low levels of perhaps 12 percent at the end of the Devonian (359 million years ago) to the highest levels in Earth's history by somewhere in the Permian (299 to 251 million years ago). The Rhynie Chert, which yielded the first abundant insect/arachnid fauna, is right at the oxygen maximum in the Devonian. Insects are then rare in the record until the near 20 percent oxygen in the Mississippian-Pennsylvanian, the time interval from 330 million to 310 million years ago—the time of the diversification of winged insects. The correspondence to the Berner curve is remarkable.

Let's formalize this relationship between oxygen levels and the first arthropod land life:

> **Hypothesis 5.1: The conquest of land by vertebrate groups was enabled by a rise in atmospheric oxygen levels during the Silurian time interval. Had atmospheric oxygen levels not risen, animals might never have colonized land.**

But we know that, following this colonization, animals became seemingly rare during the time of low oxygen. There are three possibilities. First, this seeming pause in the colonization of land was not real at all—it is simply an artifact of a very poor fossil record for the time interval from 400 million to about 370 million years ago. Second, the pause was real—that because of very low oxygen there were indeed very few arthropods, and especially insects, on land. But the few that survived were able to diversify into a wave of new forms when oxygen again rose, some 30 million years later. Third, the first wave of attackers coming from the sea as part of the invasion of land were wiped out in the oxygen fall. Yes, here and there, a few survivors held out. But the second wave was just that—coming from new stocks of invaders, again swarming onto the land under a new curtain of oxygen.

Which of these three possibilities is the correct one? The answer now seems clear. As I wrote this book, it seemed that both land arthropods and land vertebrates showed the same pattern, which could not simply be a coincidence. With help from three colleagues—Robert Berner, Conrad Labandeira (the world's foremost expert on early land arthropods), and Michel Laurin (who provided data on early land vertebrates)—I wrote a scientific paper suggesting the two-part colonization. Let's formalize this as part of our revisionist history:

*Hypothesis: 5.2 The colonization of land by animals (arthropods and, as we shall see, vertebrates as well) took place in two distinct waves: one from 430-410 million years ago, the other from 370 onward.*

Arthropods were not the only colonists to make a new life on land, of course. Gastropod mollusks also made the evolutionary leap onto land but not until the Pennsylvanian (thus they were part of the second wave), when oxygen levels were even higher than at any time during the first wave. The very inefficient gastropod lung required this high oxygen, especially in the transitional phase, when the first lungs in these formally aquatic animals were being formed. Another group that made it ashore was horseshoe crabs, at about the same time that the mollusks landed. But these are minor colonists compared to the group that most concerns this history of life—our group—the vertebrates.

## THE EVOLUTION OF TERRESTRIAL VERTEBRATES

Let's now turn our attention to the evolution of the first amphibians, the vertebrate group that first colonized the land, or partially did. The fossil record has given us a fair understanding of both the species involved in this transition and the time. A group of Devonian Period bony fish known as Rhipidistians appear to have been the ancestors of the first amphibians. These fish were dominant predators, and most or all appear to have been fresh water animals. This in itself is interesting and suggests that the bridge to land was first through freshwater. The same may have been true for the arthropods as well.

The Rhipidistians were seemingly preadapted to evolving limbs capable of providing locomotion on land by having fleshy lobes on their fins. The still-living coelacanth provides a glorious example of both a living fossil and a model for envisioning the kind of animal that did give rise to the amphibians. But another group of lobe-finned fish, the lungfish, also is useful in understanding the transition, not in terms of locomotion but in the all-important transition from gill to lung. The best limbs in the world were of no use if the amphibian-in-waiting could not breathe. There were thus two lineages of lobe-finned fishes, the crossopterygians (of which the coelacanth is a member) and the lungfish.

There is controversy about which of these groups was the real ancestor of the amphibians. Whichever it was, there is a record of the first "tetrapods," animals with four legs, in the latter part of the Devonian, meaning that the crucial transition from a fish with lobed fins and gills to an animal with four legs took place prior to that time. But when? And just how terrestrial *were* those first tetrapods? Could they walk on land? More importantly, could they breathe in air without the help of water-breathing gills as well? Both genetic information and the fossil record are of use here. But in some ways we are very hampered. Not until we somehow find the earliest tetrapods with fossil soft parts preserved will we be able to answer the respiration question.

Happily we have extant representatives of the crossopterygians and lung fish, and some relatively primitive amphibians. Geneticist Blair Hedges has compared their genetic codes in an effort to discover the time that fish and amphibians diverged. The "molecular clock" discoveries seem to roughly match the fossil record. The split of the amphibian stocks from their ray-finned ancestors (in this case the lobe fins) is dated at 450 million years ago, or at about the transition from the Ordovician Period to the Silurian Period. But this may have simply been the evolution of the stock of fish from which the amphibians ultimately came, not the amphibians themselves.

Paleontologist Robert Carroll, whose specialty is the transition of fish to amphibians, considers that a fish genus known as *Osteolepis* is the best candidate for the last fish ancestor of the first amphibian, and this fish genus did not appear until the early to middle part of the Devonian, or, that is, this final fish ancestor did not appear before about 400 million years ago. However, the first land-dwelling amphibians may have evolved 10 million years before this time, based on tantalizing evidence from footprints recently found in Ireland. A set of footprints from Valentia, Ireland, has been interpreted as the oldest record of limbed animals leaving footprints. But were these footprint makers really on land—or were they water-breathing fish that had evolved four legs to gently pad across the muddy bottom of ponds, as suggested by amphibian expert Michel Laurin in a letter to me in 2006? There are no skeletons associated with this track way, which is composed of about 150 individual footprints of an animal walking across ancient mud dragging a thick tail. This find has set off intense debate, since it predates the first undoubted tetrapod *bones* by 32 million years! The footprints were found at a time interval when oxygen levels either approached or exceeded current levels, and it is at this same time that the fossil record of insects, recounted above, yielded the first specimens of terrestrial insects and arachnids. Thus, just as high oxygen aided the transition from water to land in insects, so too might it have allowed evolution of a first vertebrate land dweller.

The first tetrapod bone fossils are not known until their appearance in rocks of about 360 million years in age, so the transition from fish to amphibians was in this interval between 400 million and 360 million years ago. A rapid drop in oxygen characterizes this interval, and the first tetrapod fossils come from a time that shows oxygen minima on the Berner curve. It is likely, however, that the actual transition from fish to amphibians must have happened much earlier, nearer the time of the Devonian high-oxygen peak but still in a period of dropping oxygen. This scenario fits the proposal that the times of low, or lowering oxygen, stimulated the most consequential evolutionary changes—the formation of new body plans, which the first tetrapod most assuredly was.

Most of our understanding about the transition from fish to amphibians comes from only a few localities, with the outcrops in Greenland being the most prolific in tetrapod remains. Although the genus *Ichythostega* is given pride of place in most discussions of animal evolution as being first, actually a different genus, named *Ventastega,* was first, at about 363 million years ago, followed in several million years by a modest radiation that included *Ichythostega, Acanthostega,* and *Hynerpeton.* Are these forms legged fish or fishy amphibians? They are certainly transitional and difficult to categorize. Of these, *Ichthyostega* is the most renowned. Its bones were first recovered in the 1930s, but they were fragmentary, and it was not until the 1950s that detailed examination led to a reconstruction of the entire skeleton. The animal certainly had well-developed legs, but it also had a fish-like tail. Nevertheless, the legs led to its coronation as the first four-legged land animal. It was only later that further study showed that this inhabitant from so long ago was probably incapable of walking on land. Newer studies of its foot and ankle seemed to suggest that it could not have supported its body without the flotation aid of being immersed in water.

The strata enclosing *Ichthyostega* and the other primitive tetrapods from Greenland came from a time interval soon after the devastating late Devonian mass extinction, whose cause was most certainly an atmospheric oxygen drop that created widespread anoxia in the seas. The appearance of *Ichthyostega* and its brethren may have been instigated by this extinction, since evolutionary novelty often follows mass extinction in response to filling empty ecological niches (the traditional view)—and since it was a time of lower oxygen (the view here). And, as postulated in this book, while periods of low oxygen seem to correlate well with times of low organism diversity, just the opposite seems true of the process bringing about radical breakthroughs in body plans: while times of low oxygen may have few species, they seem to show high disparity—the number of different body plans. Such was the case during the Cambrian Explosion, a time of relatively low species-level diversity but of many kinds of body plans relative to the number of species. So too with the interval of time from 365 million to perhaps 360 million years ago, with many new evolutionary experiments being tried out. *Ichthyostega* was one of these, and, judging from its geological record, a not too successful one. The fossil record shows that soon after its first appearance, it and the other pioneering tetrapods disappeared.

But were *Ichthyostega* and the two or three allied forms found with it even land-dwelling organisms? The bones of this first amphibian have been reexamined in detail by Cambridge paleontologist Jenny Clack. What she and other anatomists discovered was unexpected. Taken together, the anatomy of *Ichthyostega* does not seem appropriate for life on land: *Ichthyostega* would have been very inefficient on land, if it could walk in air at all. This creature was pretty much a fish with legs, rather than an amphibian in the sense of how we know them today. And if it were the first amphibian, we would expect this great evolutionary breakthrough to be soon followed by an adaptive radiation, the rapid proliferation of new species using the breakthrough morphology. But this did not happen. There was a long gap before more amphibians appeared. This gap has perplexed generations of paleontologists and it came to be known as Romer's Gap, after the early twentieth-century paleontologist Alfred Romer, who first brought attention to it. The expected evolutionary radiation of amphibians did not take place until about 340 million to 330 million years ago, making Romer's Gap at least 20 million years in length. This radiation took place at a time when oxygen had again risen to, or above, present-day levels, and that did not happen until later than 355 million years ago. A 2004 summary by John Long and Malcolm Gordon similarly interpreted the tetrapods living 370 million to 355 million years ago, the time of a great oxygen drop, as entirely aquatic—essentially fish with legs—even though some of them had lost gills. Respiration took place by gulping air, in the manner of many current fish, and by oxygen absorption through the skin. None were amphibians as we know them today, species that can live their entire adult lives on land. And it appears that none of the Devonian tetrapods had any sort of tadpole stage; they went directly from egg to land dweller without a water-breathing larval stage.

## PLUGGING ROMER'S GAP?

The long interval supposedly without amphibians was "plugged" in 2003 by Jenny Clack with great media fanfare. While looking through old museum collections she came upon a fossil long thought to have been a fish. But more detailed examination showed it to be a tetrapod and, more than that, it was an animal with five toes and the skeletal architecture that would have allowed land life. More importantly, it was within the mysterious Romer's Gap. The popular press reports of this finding, which was named *Perdepes,* would have us believe that Romer's Gap was filled. Hardly. *Perdepes* may indeed have been the first true amphibian, and it did come from an interval of time within the gap: the fossil comes from the time interval between 354 million and 344 million years ago. But here is where reality sometimes escapes the news. Dating sedimentary rocks is devilishly hard. And more so for non-marine deposits. *Perdepes* was not an amphibian living the 10 million years from 354 to 344 million years ago. Instead, it is an admittedly (by its discoverer) short-ranging genus living sometime in that interval. *Perdepes* does not plug the gap—it is a small boat sailing in a vast sea of time. It does tell us that somewhere in the middle of Romer's Gap a tetrapod did evolve the legs necessary for land life. But did it breathe air? Could it live entirely emerged from the water, even for a few minutes? That we do not know. So let's demystify the gap, as alluded to earlier in this chapter.

Alfred Romer thought that evolution of the first amphibians came about because of the effect of oxygen. But the pathway may not have been that supposed by Romer, who considered that lungfish or their Devonian equivalents were trapped in small pools that would seasonally dry up. In his view the lack of oxygen brought about by natural processes in these pools, and the drying, was the evolutionary impetus for the evolution of lungs. His idea was that seasonally drying swamps spurred the jump to land or smaller freshwater ponds or lakes. According to this idea, then, the amphibians-in-waiting were forced out of these pools and into air. Gradually, those animals that could survive the times of emersion from water had an advantage. These fish still had gills, but the gills themselves allowed some adsorption of oxygen. The problem was that the gills quickly dried out. By evolving ever-tighter and water-resistant pockets around the gills, the transition from gill to lung was under way. But a gill is still an evagination, even if in a pocket. There had to be a complete inversion of this system, for a lung is a series of sacs, whereas a gill is a series of protuberances. It may be that the transitional forms had both gills and primitive lungs.

The transition from aquatic tetrapods such as *Ichthyostega* or, more probably, *Perdepes,* involved changes in the wrists, ankles, backbone, and other portions of the axial skeleton that facilitate breathing and locomotion. Rib cages are important to house lungs, while the demands of supporting a heavy body in air, as compared to the near flotation of the same body in water, required extensive changes to the shoulder girdle, pelvic region, and the soft tissues that integrated them. The first forms that had made all of these changes can be thought of as the first terrestrial amphibians and the *Perdepes*, found in rocks younger than 355 million years in age, may indeed have been the first of all, according to Long and Gordon. But there may be a continuation of the gap after Perdepes. The great radiation of new amphibian species did not occur until 340 million to 330 million years ago. But when it finally took off, it did so in spectacular fashion, and by the end of the Mississippian Period (some 318 million years ago) there were numerous amphibians from localities all over the world.

Let's reexamine the radiation of amphibian species in the context of atmospheric oxygen levels. A gill is very inefficient when it must act as a lung, and a primitive lung must evolve through many steps before the complex and high surface area surface of internal pockets, all vascularized, with concomitant changes to the circulatory system, is effected. While all these respiratory and circulatory changes are happening, the respiratory system in the stock leading to amphibians would have been less efficient at delivering oxygen than either a gill in water or the complex lung in air that would later be evolved. The high-oxygen peak in the early Devonian would have provided the extra oxy-

*Reconstructions of the earliest known tetrapods,* Tiktaalik *(left) and* Acanthostega *(right), shows how the transition from fish to amphibians took place. In spite of their limbs, both of these were probably fully aquatic and unable to climb onto land because of inadequate (for land life) respiratory and locomotory systems.*

gen necessary to make this system work, as would have the high oxygen of the latter parts of the Mississippian and Pennsylvanian of 330 million to 300 million years ago.

The Berner curve starting this chapter suggests that there was a great drop in oxygen near the end of the Devonian and coincident with this is the Devonian mass extinction, one of the five most severe mass extinctions in Earth's history. While investigators have been searching for clues to this extinction for decades and have invoked causes ranging from an asteroid impact to climate change, it is not known for sure what the causes of the Devonian mass extinction were. Ammonite workers have long known that at this time the oceans showed a marked change to low-oxygen conditions. The extinction took place over about 2 million years, from 370 to 368 million years ago, at a time when the Berner curve shows a very low level of atmospheric oxygen of about 12 to 14 percent.

Here is where the new terrestrial arthropod data from Conrad Labandeira and the new land vertebrate range data from Michel Laurin can help solve the mystery of "Romer's Gap," and support the hypothesis presented below that the animal conquest of land happened in two initial phases separated by a time of low oxygen. The figure from our paper is shown here:

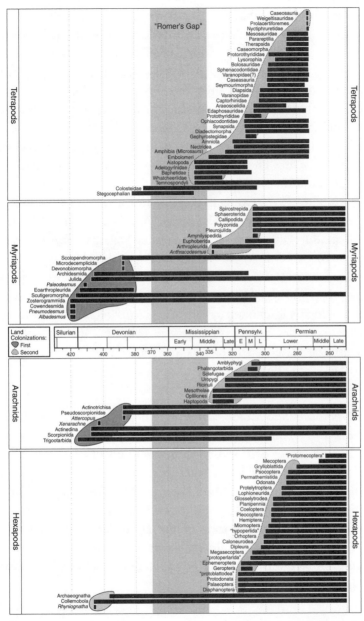

The black lines mark the geological ranges of arthropods and vertebrates during the crucial time when land animals first invaded the land. The grey shaded region is the time known as Romer's Gap. These data indicate that the conquest of land was made up of two separate events, with the first largely a failure.

*Ichthyostega*, long thought to mark the appearance of the first land vertebrate, may have been far more fish-like than first thought, and the fact that it lost its gills is not evidence of a fully terrestrial habitat. Long and Gordon point out that today there is a large diversity (over a hundred different kinds) of air-breathing fish. Air breathing has evolved independently in as many as 68 kinds of these extant fish, showing how readily this adaptation can take place. *Ichthyostega* may not even have been on the line leading to the rest of the tetrapod lineages but may have been on a line that was evolving back into a fully aquatic life style, forced off the land by its primitive lungs and the dropping oxygen levels of the late Devonian. It may even be a descendent of the first true tetrapods, perhaps evidenced by the Valentia footprints of the early Devonian. But while there is doubt that the Valentia footprints came from the first land tetrapod, there is no doubt that the first really diverse land animal fauna, dominated by air-breathing arthropods, appear in the fossil record coincident with the early Devonian oxygen high. This high was followed by a low-oxygen period, when *Ichthyostega* appeared and then quickly disappeared. Following the extinction of the *Ichthyostega*, the world had to wait until oxygen again increased before land could be conquered.

Let's thus formalize this view:

*Hypothesis 5.3: Colonization of the land came in two steps, each corresponding with a time of high oxygen: the first invasion was from 410 million to 400 million years ago and was followed by a second one, beginning from 370 million years ago, that dramatically increased the diversity of land animals with the oxygen high of 330 million to 300 million years ago. The time in between the time of the Devonian mass extinction through the so-called Romer's Gap had little animal life on land. Romer's Gap should be expanded in concept to include arthropods and chordates.*

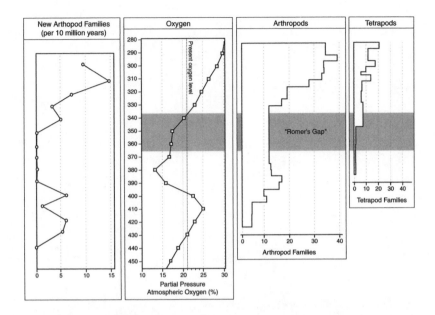

*A comparison of oxygen levels with the diversity of arthropods and limbed vertebrates during the interval of time shown in the previous figure. The far left column, which portrays the number of new terrestrial arthropod orders over the time interval from 450 million to 290 million years ago shows that during the time from 380 million to 350 million years ago, a time interval encompassing Romer's Gap, there were no new orders. This time interval also coincides with either dropping- or low-oxygen values.*

## THE TIME OF GIANTS

Romer's Gap ended in the Carboniferous Period (split in two in America, where we call it the Mississippian and Pennsylvanian Periods), and its European name comes from the fact that the majority of coal now found on Earth dates from this time. During this time oxygen levels rose in spectacular fashion, and in the last intervals of the Carboniferous and continuing into the successive Permian Period, oxygen levels finally topped out at nearly 35 percent, creating a unique interval in Earth's history, the subject of the next chapter.

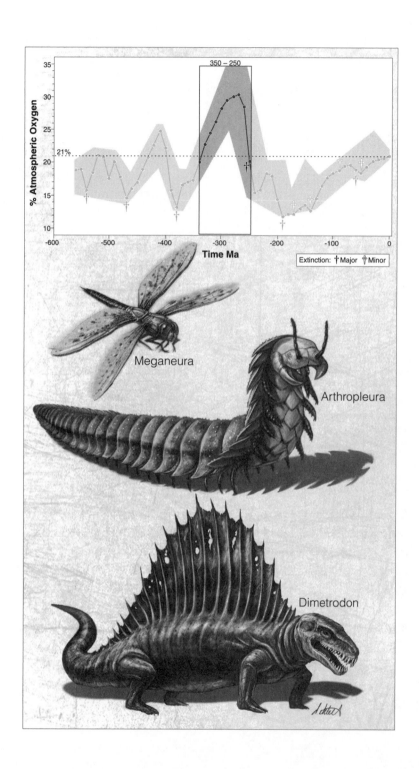

Meganeura

Arthropleura

Dimetrodon

Extinction: † Major  † Minor

## 6

### THE CARBONIFEROUS-EARLY PERMIAN:
### HIGH OXYGEN, FIRES, AND GIANTS

One of the most useful exercises in trying to understand the nature of body plans is to look at their limitations. For example, no echinoderm ever made it into fresh water, let alone land, and the reasons for this are not hard to find: the evolutionary move from salty to fresh water requires extensive modifications of excretory systems and salt regulation. Echinoderms apparently could never produce these changes. Yet another way to chart limitations is not by what environments were exploited (or not) by specific body plans but by asking a deceptively simple question: what is the largest size that a specific lineage attained? Since large size is often a protection against predation, it seems that most animals grow as large as they can. But there are costs to this, and, as we have seen here, one of the most stringent is in assuring that the new, bigger body receives sufficient oxygenation.

What ultimately limits this growth can provide illuminating answers about the biomechanics of animal design. A favorite example of this exercise deals with arthropods. A spate of wonderful science fiction "B" movies from the 1950s (and then sporadically up to the present-day, such as several years ago when giant mosquitoes sucked unto death a bunch of campers) unleashed a striking diversity of giant insects and spiders. And some were gigantic indeed. The ants of *Them!* were the size of tanks, while the deadly mantis was the size of a 747. Yet we know that insects, with the external body armor typical of arthropods, could never live at such size. Because of scaling properties and strength of the chitin making up the arthropod exoskeleton, a giant ant or preying mantis of even human size would collapse, its walking legs snapping. But there is another aspect of arthropod design that limits size—respiration. Insects, spiders, and scorpions appear to be size-limited by the degree to which oxygen can diffuse into the innermost regions of their bodies. Today, no insect is bigger than about 6 inches in body length (although arthropods in water can be and are bigger, since their weight is buoyed up by the watery medium they exist in). In the past, however, much larger land arthropods than this did exist—during the interval of the highest oxygen in Earth's history, the subject of this chapter. Here we will explore how the highest oxygen in Earth's history allowed some of the strangest—and largest— animals ever on Earth.

## VOYAGE

Here is what we might see were we to return to the world of 300 million years ago, when oxygen was at its highest in Earth's history. The first noticeable characteristic is the color of the sky. It is a polluted yellow-brown, irrespective of weather; only in high winds does the air clear and then soon it muddies again. This is due to smoke from giant fires perpetually raging and new ones set alight with each lightning strike hitting the extensive forests of the temperate and tropical regions. But there is more than soot in the air; there is fine wind-blown dust as well, accumulating as loess, or glacial front deposits. The air is cold over much of the planet, for this is the time of one of the most extensive glaciations in Earth's history, with ice caps at each pole and continental glaciers reaching icy fingers across the land as they snake downward out of the mountains onto the plains and river valleys. Where there are forests we find unending vistas of conifer-like trees, for the gymnosperms have by now evolved and have swept away many of the earlier dominant plants. No longer do tall but shallow-rooted primitive trees like *Lepidodendron* dot the landscape. It is much more familiar, or familiar at least to those of us who now live in the high northern or southern latitudes where pines and fir trees dominate the forests.

Unlike during any of our previous trips, the land is alive with animals and the arthropods, at least, are giants relative to similar species living today. Enormous insects, including dragonflies with nearly yardwide wings, flit about above swampy forests, and even the drier upland areas show a high diversity of both flying and earth-bound insects, intermingled with spiders, scorpions, and millipedes, and many are also giants. But it is the vertebrate life that interests us most, and here too there are giants, at least compared to the very first terrestrial forms of some 40 million years earlier, in the early part of the Carboniferous (but certainly not giant relative to the dinosaurs of the future). In the swamps, giant amphibians, some 10 feet long and massive in girth, lie about like modern-day crocodiles. Land vertebrates, freed from water by the evolution of amniotic eggs, are similarly large. Chief among them is the Sail Back, *Dimetrodon,* and among its pack scurry smaller reptiles, the eventual rootstocks of turtles, lizards, crocodiles, and dinosaurs.

We turn to the oceans, where a huge diversity of fish is immediately noticeable. The largest are the cartilaginous sharks, some with peculiar heads and teeth, but bony fish are there too. Missing are the placoderms and ostracoderms. Chambered cephalopods compete for dominance with the smaller fish, but gone are the enormous straight nautiloids of earlier times. Almost all nautiloids are gone, in fact, replaced by ammonoids—cephalopods with shells that look like that of the living chambered nautilus but showing soft parts more akin to modern squids than the archaic soft parts of the nautilus. Yet as interesting as the active swimmers are, it is the nature of the attached animals that attracts immediate interest. Moving in shallow water above a wide and warm continental shelf we find seemingly endless fields of flowers. But flowers will not appear on Earth for another 200 million years. A closer look shows only the resemblance of these long-stalked animals with petal-like heads to the flowers we know. They are crinoids, or sea lilies, animals banished in our world to the marginal habitats of the less successful. But there is clearly no lack of success here, and the appearance of square miles of these fantastic stalked echinoderms is puzzling. We have seen them before in earlier oceans, but not like this. Virtually alone among echinoderms, crinoids have no respiratory structures, but the oxygen levels of this sea are higher than at any time in Earth's history, and this change has led to the ascendance of crinoids. They are so abundant that their stony stalks make up the majority of bottom sediment, for most are relatively large for their kind.

We pass to a reef habitat, and once again the changes are evident. Sponges have taken the place of corals as the major framework builders of the reef ecosystems. Some of these reefs extend along the shelf-slope break for hundreds of miles, a habitat filled with fish and ammonoids among the giant sponges. Large as well are resident brachiopods, with two kinds in abundance: one with spines, the other like a large garbage can sitting among the other reef organisms. Both are huge for their race—testament again to the high oxygen of this world.

### INSECTS IN THE CARBONIFEROUS-PERMIAN OXYGEN HIGH

Oxygen reached extraordinarily high values in the time interval from about 320 million to 260 million years ago, with maximal values occurring near the end of this interval. The Carboniferous Period (in North America subdivided into the Mississippian and Pennsylvanian Periods) and the first half of the subsequent Permian Period (299 million to 251 million years ago) were the times of high oxygen, and the biota of the world at that time left clear evidence of the high oxygen. Insects from that time present the best evidence.

The Carboniferous high (and much else as well) was wonderfully described by Nick Lane in his 2002 book, *Oxygen*. Lane wrote about a fossil dragonfly discovered in 1979 that had a wingspan of some 20 inches. An even larger form with a 30-inch wingspan is known from fossils of this Carboniferous time—a beast aptly named *Meganeura*, yet another dragonfly (*Mega* means large). And it was not only the wings that were large. The bodies of these giants were proportionally larger, with a width of as much as an inch and a length of nearly a foot. This is about seagull size and while seagulls are never linked in any sentence with the word "giant," an insect with a 20-inch wingspan was indeed a veritable giant. In comparison, today's dragonflies may reach a 6-inch wingspan but far more commonly are much smaller. Other giants of the time included mayflies with 19-inch wingspans, a spider with 18-inch legs, and yard-long (or longer) millipedes and scorpions. A 3-foot-long scorpion could weigh 50 pounds and would be a formidable predator of all land animals, including the amphibians. But as we will see, the amphibians also evolved some giant species of their own.

So is it the biomechanics of legs that dictated (or limited) insect size? Apparently not. It is the insect respiratory system that dictates maximum size, it seems. All insects use a system of fine tubes called trachea. Air diffuses into the tissues from these tubes, and air is actively ventilated into the tubes. Either by the insect's rhythmic expansion and contraction of the abdominal region, or by the insect's flapping of its wings to create air currents around the tracheal opening, air is pulled into the canals. The tracheal system is fantastically efficient in either case. Flying insects achieve the highest metabolic rates of any animal, and experimental evidence shows that increasing oxygen to higher levels enables dragonflies to produce even higher metabolic rates. These studies showed that dragonflies are both metabolically limited and probably size limited as well by our current 21 percent oxygen levels.

Whether or not oxygen levels control arthropod size has been contentious. The best evidence that it does comes from studies of amphipods, small marine arthropods that are widely distributed in our world's oceans and lakes. Gautier Chapelle and Lloyd Peck examined 2,000 specimens from a wide variety of habitats and discovered that bodies of water with higher dissolved oxygen content had larger amphipods. More direct experiments were conducted by Robert Dudley of Arizona State University, who grew fruit flies in elevated oxygen conditions and discovered that each successive generation was larger than the preceding when raised at 23 percent oxygen. In insects, at least, higher oxygen very quickly promotes larger size.

It was not only higher oxygen that allowed the existence of giant dragonflies. The actual air pressure is presumed to have been higher too. Oxygen partial pressures rose but not at the expense of other gases. The total gas pressure was higher than today, and the larger number of gas molecules in the atmosphere would have given more lift to the giants. Gas pressure is a function of the number of molecules in the air.

The oxygen high happened when more and more oxygen molecules entered the atmosphere. But this addition did not cause the number of nitrogen molecules to become reduced—hence, higher atmospheric gas pressure.

Why so much oxygen in the air then? In Chapter 2 we saw that oxygen levels are mainly affected by burial rates of reduced carbon and sulfur-bearing minerals like fool's gold (pyrite). When a great deal of organic matter is buried, oxygen levels go up. If this is true, it must mean that the Carboniferous period, the time of Earth's highest oxygen content, must have been a time of rapid burial of large volumes of carbon and pyrite, and evidence from the stratigraphic record confirms that this indeed happened—through the formation of coal deposits.

We are looking at a long interval of time—70 million years, longer than the time between the last dinosaurs and the present-day—in the 330 to 260 million years of high oxygen. It turns out that 90 percent of Earth's present-day coal deposits are found in rocks of that interval. The rate of coal burial was thus much higher than during any other time in Earth's history—600 times higher, in fact, according to Nick Lane in his book *Oxygen*. But the term "coal burial" is pretty inaccurate. Coal is the remains of ancient wood, and thus we see a time when enormous quantities of fallen wood were rapidly buried and only later through heat and pressure turned to coal. The Carboniferous Period was the time of forest burial on a spectacular scale.

The burial of organic material during the Carboniferous was not restricted to land plants. There is much carbon in the oceans tied up in phyto- and zooplankton, the oceanic equivalents of the terrestrial forests, and here too large amounts of organic-rich sediments accumulated on sea bottoms.

The ultimate cause of this unique accumulation of carbon, leading to the unique maximum of oxygen levels, was the coincidence of several geological and biological events. First, the continents of the time coalesced into one single large continent by the closing of an ancient Atlantic Ocean. As Europe collided with North America and South America with Africa, a gigantic linear mountain chain arose along the suturing of these continental blocks. On either side of this mountain chain great floodplains arose, and the configuration of the mountains produced a wet climate over much of the planet. Newly evolved trees colonized the vast swamps and the adjoining drying land areas that came into being. Many of these trees would appear fantastic to us in their strangeness, and one of their strangest traits was a very shallow root system. They grew tall and fell over quite easily. There are lots of falling trees in our world but nowhere near the accumulation of carbon.

More was at work than a swampy world ideal for plant growth. The forests that came into being some 375 million years ago were composed of the first true trees that used lignin and cellulose for skeletal support. Lignin is a very tough substance, and today it is broken down by a variety of bacteria. But even after nearly 400 million years—which brings us to the modern-day—the bacteria that do this job take their own sweet time. A fallen tree takes many years to "rot," and some of the harder woods, those with more lignin than the so-called soft woods like cedar and pine, take longer yet. This "rotting" is accomplished by the oxidation of much of the tree's carbon, so even if the end product is eventually buried, very little reduced carbon makes it into the geological record. Reduced carbon, as we saw in Chapter 2, is carbon that is deposited in the absence of oxygen in a state that is highly reactive when and if it is later exposed to free oxygen. Back in the Carboniferous many, or perhaps all, of the bacteria that decompose wood were not yet present. Trees would fall and not decompose. Eventually sediment would cover the unrotted trees and reduced carbon would be buried in the process. With all of these trees (and the plankton in the seas) producing oxygen through photosynthesis and very little of this new oxygen being used to decompose the rapidly growing and falling forests, oxygen levels began to rise.

## OXYGEN AND FOREST FIRES

The Carboniferous oxygen peak would have had consequences in addition to gigantism. Oxygen is combustible, and the more there is the bigger the fire—it facilitates fuel ignition, and the fuel in question was the huge and global forest of the Coal Age. The Carboniferous Period may have witnessed the largest forest fires ever to occur on Earth—at least until the dinosaur-killing and forest-igniting Chicxulub asteroid strike (the well known Cretaceous-Tertiary, or "K/T" extinction event) of 65 million years ago. Like so much dealing with the change of oxygen over time, studies on the possibility of megafires provoked by high atmospheric oxygen have been controversial but are becoming much less so as more and more evidence accumulates. Indeed, the forest fire controversy has prompted a major criticism of the theory that oxygen values were different in the past (including higher). It was suggested that ancient forests would not have been able to survive the catastrophic fires, and since we have a long fossil record of the forest, the implication is that catastrophic fires did not take place. Conditions of elevated oxygen, at least theoretically, should generate more rapid rates of flame spreading, and higher-intensity fires and indeed large deposits of fossil charcoal in sedimentary rocks of Mississippian and Pennsylvanian age in North America are evidence that there were forest fires back then—forest fires that were larger, more frequent, and more intense than those of today, although direct comparison suffers from the very different biological makeup of forests then and now.

If there were more and more intense forest fires, we would expect to see morphological adaptations to promote fire resistance over time. Plants evolved a series of adaptations collectively known as fire resistance traits, which include thicker bark, deeply imbedded vascular tissue (Cambria), and sheaths of fibrous roots surrounding the stems.

So why didn't the Carboniferous forest burn to the ground? While fires seem to have been more frequent then, the presence of fire-resistant plants and high moisture content both in the plants themselves and in the swampy terrain of large portions of Earth's surface in the numerous coal swamps limited damage.

## THE EFFECT OF HIGH OXYGEN ON PLANTS

So far we have concentrated on the effects that varying levels of oxygen have on animals. But animal life is itself totally dependent on plant productivity, and we are sure that major perturbations in plant diversity and/or abundance had effects on the animal food chains. Thus, we should look at the effects of varying oxygen levels on plants.

Like animals, plants need oxygen for life. Oxygen is taken up within the cells during photorespiration. But the levels are far lower than those needed by animals for the most part, and a second difference is that various parts of terrestrial plants, for instance, have different oxygen needs. Compare the very different environments of underground roots (surrounded by water, both solid and gas, but without light) to leaves (surrounded by gas only, with intermittent light). It is the root system that is most susceptible to damage or cell death from low-oxygen conditions, and it is also the underground environment where such conditions can occur even at times of well-oxygenated air. Roots can be smothered by ground water with low-oxygen values, for instance.

What about plants and high-oxygen levels? Here there are far fewer data, but what is known suggests that elevated levels of oxygen are deleterious to plants. Higher levels of oxygen in air lead to increased rates of photorespiration, but a more serious consequence is that with higher oxygen levels there are more toxic hydroxyl radicals that are dangerous to living cells, including all animals. To further test these possibilities, David Beerling of Yale University grew various plants in higher than current oxygen within closed tanks. When oxygen levels were raised to 35 percent (thought to have been the highest levels of all time, occurring in the late Carboniferous or early Permian), net primary productivity (a measure of plant growth) dropped by one-fifth. It may be that the higher oxygen of the Carboniferous-early Permian caused a reduction in plant life to some degree, although this is not observable in the fossil record by any dramatic change or mass extinction during this interval.

## SLUGGISH EVOLUTION

Throughout this book, it is proposed that times of low oxygen in Earth's history stimulated new kinds of evolution but were also times of low diversity. During high-oxygen times, conversely, diversity is high but the rate of new species formation is low. This hypothesis is based on the proposal that low oxygen forces new experimentation in terms of body plan. This proposal is supported by a new comparison of oxygen levels with data on the rate of new taxon formation. Thus, we should expect to see a very low rate of new species formation during the oxygen high of the Carboniferous-early Permian. Just such a finding has recently been made. In 2005, paleontologist Matthew Powell of Johns Hopkins University compiled voluminous data on the fates of various marine invertebrates during this oxygen high. He discovered very low rates of both origination and extinction. In other words, few new taxa appeared, and those that were already present rarely went extinct. The marine world was composed of an assemblage of virtual living fossils, which are characterized by long ranges (they existed for long periods of time) and produced very few new species.

Why did this happen? Powell invoked the presence of the late Paleozoic glaciation as the cause:

> The Paleozoic history of marine life was interrupted in late Paleozoic time by a conspicuous interval of sluggish diversification and low taxonomic turnover. This unique interval coincided precisely with the late Paleozoic Ice Age.

Powell went on to suggest that the cool climate was the cause of this slow interval of evolution. Yet other times of glaciation seem to contradict this. One of the greatest extinctions of all time, that of the Ordovician (the mass extinction discussed in Chapter 4), has been blamed on the glaciation by most experts, and noted paleontologist Steve Stanley has suggested that cooling is a killer and cause of mass extinction. In our different view the sluggish evolution seen during the late Paleozoic is related to the high level of oxygen.

So how did all of this relate to the group we belong to and the group most people are interested in—the vertebrates?

## OXYGEN AND LAND ANIMALS—REPTILES AND THEIR EGGS

Conquest of the land by chordates, our lineage, required many major adaptations. The most pressing was a way to reproduce that allowed development of the embryo in an egg out of water. The amphibians of the Pennsylvanian and Permian presumably still laid eggs in water, and thus they could not exploit the resources of land regions that were without lakes or rivers. The evolution of what is termed the amniotic egg solved this, and it was this egg that ensured the existence of a stock of vertebrates now known as reptiles. The evolution of the amniotic egg differentiates the reptiles, birds, and mammals from their ancestral group, the amphibians. Before the end of the Mississippian Period, three great stocks of reptiles had diverged from one another to become independent groups: one that gave rise to mammals, a second to turtles, and a third to the other reptilian groups and to the birds. The fossil record shows that there are many individual species making up these three. A relatively rich fossil record has delineated the evolutionary pathway of these groups. It has also required a reevaluation of just what a "reptile" is. As customarily defined, the class Reptilia includes the living turtles, squamates, and crocodiles. Technically, reptiles can now be defined by what they are not: they are amniotes that lack the specialized characters of birds and mammals. The fossil record suggests that the "amniotes" are "monophyletic"—that they have but one common ancestor—rather than suggesting the possibility that amniotic eggs arose independently more than once.

Reptiles are considered to be a monophyletic stock that diverged from amphibian ancestors perhaps sometime in the Mississippian Period of more than 320 million years ago. But while genetic evidence of this divergence, obtained by the "molecular clock" analysis technique, can be dated back to as long ago as 340 million years, fossils that are ascribed to the first reptiles (instead of terrestrial amphibians) have been recovered from several localities globally. Fossils of small reptiles named *Hylonomus* and *Paleothyris* have been found interred in fossilized tree stumps of early Pennsylvanian age, and it may be that the fossil record of this later appearance is more valid than the assumption of a Mississippian evolution of the group. In either case, these first reptiles were very small—only about 4 to 6 inches long.

That these small reptiles laid the first amniotic eggs is still speculation. There are no fossil eggs in the stratigraphic record until the lower Permian, and this single find remains controversial. But the pathway to the amniotic condition probably passed through an amphibian-like egg without a membrane that would reduce desiccation, which laid in moist places on land. It would have been the evolution of a series of membranes surrounding the embryo (the chorion and amnion), covered by either a leathery or a calcareous but porous egg that was required for fully terrestrial reproduction. One possibility seemingly never considered is that these first tetrapods evolved live births, so that the embryos were not born until substantial development within the female had taken place; we would love to know if any animals of this time produced live births, but that is only rarely recorded in the fossil record. One exception is the extraordinary fossil showing a female ichthyosaur of the Jurassic that died giving live birth. We also have no record of eggs laid in water—for instance, frog and salamander eggs are so soft that they are never preserved.

Land eggs eventually were produced, and it was here that the level of oxygen must have played a part. There is a huge tradeoff in reproduction for any land animal using an egg-laying strategy. Moisture must be conserved, so the openings of the egg must be few and small. But reducing permeability of the egg to water moving from inside to outside also reduces the movement of oxygen into the egg by diffusion. Without oxygen the egg cannot develop. The first amniotic eggs were probably produced in oxygen levels equal to or even higher than those of today. It may be no accident that the evolution of the first amniote occurred during a time of high oxygen. This leads us to a new hypothesis:

**Hypothesis 6.1: Reproductive strategy is affected by atmospheric oxygen content, with higher-oxygen contents producing more rapid embryonic development. High oxygen may have allowed amniotic eggs and then live births.**

Some biologists have suggested that live births could not take place in low oxygen because, at least in mammals, the placenta delivers lower levels of oxygen than are present even in arterial blood in the same mother. But this is for mammals. Reptiles have a very different reproductive anatomy. It may be that low oxygen even favors live birth. Evidence to support this comes from three lines of evidence. First, birds (egg layers) living in high-altitude habitats routinely feed at higher altitudes than the maximum altitude at which they can reproduce. The maximum altitudes of the nests of many mountainous bird species repeatedly show this pattern. The highest nests are at 18,000 feet, and higher than this the embryos will not develop successfully. While at least three factors may be involved in this limit (lowered-oxygen content with altitude, desiccation because of air dryness at altitude, and relatively low temperatures at greater altitude), it may be oxygen content that is most important. Second, recent experiments by John VandenBrooks from Yale University have shown that alligator eggs taken from natural clutches and raised in artificially higher-oxygen levels showed dramatically faster than normal development rates. The embryos grew some 25 percent faster than controls held at normal atmospheric oxygen levels. Increased oxygen clearly influences growth rates, at least in American alligators. Since egg contents are a tasty snack for many predators now, and surely back then, it is in the embryo's better survival interest to develop and hatch as quickly as possible. Finally, a higher proportion of reptiles at high altitudes use live births than do reptiles at lower altitudes.

At this point we can summarize and discuss the variables in land animal reproduction and try to relate these to generalizations about both oxygen levels and temperature. There are two possible strategies, egg laying or live birth. In the egg case, the eggs are either covered with a calcareous shell or a softer, more leathery shell. Today, all birds utilize calcareous eggs, while all living reptiles that lay eggs use the leathery covering. Unfortunately, there is little information about the relative oxygen diffusion rates for leathery, or parchment, eggs compared to calcareous eggs.

The utilization of egg laying or live births has important consequences for land animals. The embryos developed by the live birth method are not endangered by temperature change, desiccation, or oxygen deprivation. But the cost is the added volume to the mother, which must invariably make her more vulnerable to predation in addition to making her need more food than would be necessary for herself alone. Egg layers are not burdened with this problem but have the tradeoff of a less safe environment—the interior of an egg outside the body—that leads to enhanced embryonic death rate through predation or lethal conditions of the external environment.

The study of oxygen levels and egg morphology or reproductive strategy is in its infancy. Obvious tests include raising eggs at both high- and low-oxygen levels, for both calcareous and leathery eggs. Also, a direct study of fossil eggs themselves would be of great interest. A prediction is that fossil eggs from times of low oxygen should show more openings than those from times of higher oxygen.

## OXYGEN AND LAND ANIMALS—REPTILES AND THEIR LOCOMOTION

As four-legged vertebrates emerged from their piscine ancestors, many new anatomical challenges had to be overcome. No longer was there water to support the animal's body; in air, both support and locomotion had to be accomplished by the four legs. An entirely new suite of shoulder and pelvic girdle designs had to evolve, along with the muscles necessary to allow locomotion. Equally daunting was the problem of acquiring sufficient oxygen to allow sustained exercise. Early tetrapods apparently used the same set of muscles for motion and for taking a breath, and they could not do both at the same time. Fish seem to have no problem with sustained exercise or with respiring during activity, suggesting that oxygen is not a limiting factor in daily activity. For land tetrapods this is not the case. The body plan of the earliest land tetrapods provided for a sprawling posture with legs splayed out to the sides of the body trunk. In walking or running with such a body plan, the trunk is twisted first to one side and then to the other in a sinuous fashion. As the left leg moves forward, the right side of the chest and the lungs within are compressed. This is reversed with the next step. The distortion of the chest makes "normal" breathing difficult to impossible—so each breath must be taken between steps. This process makes it impossible for the animal to take a breath when running— modern amphibians and most reptiles cannot run and breathe at the same time (the exceptions including varanids that augment respiration with gular pumping), and it is a good bet that their Paleozoic ancestors were similarly impaired. Because of this there are no reptilian marathoners and not too many long-distance sprinters. This is why reptiles and amphibians are ambush predators. They do not run down their prey. The best of the modern reptiles in terms of running is the Komodo dragon, which will sprint for no more than 30 feet when attacking prey. This is called Carrier's Constraint, after it discoverer, physiologist David Carrier.

The dilemma of not being able to breathe and move rapidly at the same time was a huge obstacle to colonizing land. The first land tetrapods would have been at a great disadvantage to even the land arthropods, such as the scorpions, for the vertebrates would have been slow and would have needed to stop constantly to take a breath. This is why oxygen levels would have been critical. Only under high-oxygen conditions would the first land vertebrates have had any chance of making a successful living on land.

One consequence of limited respiration while moving was that the early amphibians and reptiles evolved a three-chambered heart. This kind of heart is found in most modern amphibians and reptiles and is adaptive for creatures that have the problem of inferior respiration while moving. While a lizard is chasing prey it is not breathing, and thus the fourth chamber of the heart, which would be pumping blood to the lungs, is superfluous. The three chambers are used to pump blood throughout the body, but the price that must be paid is that it takes the lizard longer to reoxygenate the blood when activity ceases.

One group of reptiles, the mammal-like reptiles or synapsids, either partially or totally solved the reptilian problem of not being able to breathe while running by changing their stance. The synapsids show an evolutionary trend of moving their legs into a position so that they were increasingly under the trunk of the body, rather than splayed out to the side as in modern lizards. This created a more upright posture and removed, or at least greatly decreased, the lung compression that accompanies the sinuous gait of lizards and salamanders. While there was still some splay of the limbs to the sides of the trunk, it was certainly less than in the first tetrapods. With the evolution of the therapsids in the middle Permian, the stance became even more upright.

## OXYGEN AND LAND ANIMALS—REPTILES AND THEIR THERMOREGULATION

Another important variable is the nature of thermoregulation—the possibilities of endothermy (warm-bloodedness), ectothermy (cold-bloodedness), and a third category that is essentially neither of the others and is associated with very large size. Warm-bloodedness (endothermy) allows animals using it to stay at a constant, warm temperature no matter how cold it gets. However, this can work against animals in very hot climates, as it is more difficult to cool off than warm up. Cold-blooded animals match the temperature around them. They are sluggish in the cold. There is a third kind of metabolism, found in animals so big that they are largely unaffected by daily swings in temperature, such as daytime and nighttime. The very large dinosaurs presumably used this system.

Other important aspects that might be related to the environmental conditions in which the various clades evolved and then lived include the presence or absence of scales, hair, and feathers. With thermoregulation pathways, the question of whether or not dinosaurs were warm-blooded has been the most discussed and most controversial of all. The fact that each of these characteristics, thermoregulatory systems and body covering, is primarily either physiological or involves body parts that only rarely leave any fossil record (such as fur) is in large part responsible for the controversies. We know that all living mammals and birds are warm-blooded, with the former having hair and the latter feathers, just as we know that all living reptiles are cold-blooded, with no hair or feathers. The status of extinct forms remains controversial. Of interest here is how oxygen concentration primarily and characteristic global temperatures secondarily affected thermoregulation or characteristic body coverings in animal stocks of the past. Let's look at each of the three stocks in more detail with this overarching question in mind.

## Diapsids

Openings are used to lighten the reptile skull, and their number (or lack of) is a convenient way of differentiating the three major stocks of "reptiles." Anapsids (ancestors of the turtles) had no major openings in their skulls; synapsids (ancestors of the mammals) had one; and diapsids (dinosaurs, crocodiles, lizards, and snakes) had two. The earliest member of the diapsids is known from the latest Pennsylvanian rocks around 305 million to 300 million years ago, and it was small in size, about 8 inches in total length. From the time of their origins until the beginning of the fall of oxygen, which probably began in earnest some 260 million years ago, in the middle and late part of the Permian period, the diapsids did little in the way of diversification or specialization. They remained small and lizard-like. They gave no indication that they would be the ancestors of the largest land animals ever to appear on Earth, in the form of the Mesozoic dinosaurs. If the time of highest oxygen stimulated insects to their greatest size, the same cannot be said of the diapsids.

The most pressing question is whether or not this group was warm-blooded.

## Anapsids

The lineage that ultimately gave rise to turtles was very successful during the late Pennsylvanian but less so into the Permian. Anapsids did evolve into giant forms, including cotylosaurs and the even larger pareiosaurs. These were armored giants, surely slow moving and herbivores that lived right until the end of the Permian. Other anapsids were small and more lizard-like. It is very likely that the gigantic size of the earlier Permian anapsids was allowed by high oxygen. All modern anapsids use ectothermy; they are cold-blooded. Presumably the ancient forms used this system as well, but that is still controversial.

## Synapsids

The third group of amniotes from this time, the synapsids, or mammal-like reptiles, are known in their most primitive form from Pennsylvanian rocks, and these ancestors of mammals had a lizard-like small shape and mode of life in all probability. It is assumed that these early synapsids were cold-blooded. They in turn gave rise to two great and largely temporally successive stocks, the pelycosaurs (or finbacks, like the early Permian *Dimetrodon*) and their successors, the therapsids (the lineage giving rise to mammals). It is this latter group that is also called the mammal-like reptiles.

Unlike the diapsids, the synapsids diversified during the oxygen high and at its peak became the largest of all land vertebrates. In the

*Reconstruction of the synapsid reptile* Dimetrodon. *The large fin was probably for thermoregulation, and thus is evidence that this group was not warm-blooded.*

latter part of the Pennsylvanian the pelycosaurs probably looked and acted like the large monitor lizards or iguanas of today. By the end of the Pennsylvanian some attained the size of today's Komodo dragon, and they may have been fearsome predators. By the beginning of the Permian Period, some 300 million years ago, the pelycosaurs made up at least 70 percent of the land vertebrate fauna. And they diversified in terms of feeding as well; three groups were found: fish eaters, meat eaters, and the first large herbivores. Both predators and prey (for this group evolved both predatory and herbivorous species) could attain a size of 12 feet in length, and some, such as *Dimetrodon,* had a large sail on its back that would have made it appear even larger.

The sail present on both carnivores and herbivores of the late Pennsylvanian and early Permian is a vital clue to the metabolism of the pelycosaurs: it was a device used to rapidly heat the animal in the morning hours. By positioning the sail so as to catch the morning sun, both predators and prey could warm their large bodies quickly, allowing rapid movement. The animal that first attained warm internal temperature would have been the winner in the game of predation or escape, and hence natural selection would have promoted these sails. But the larger clue from the presence of sails is that during the oxygen high, the ancestors of the mammals had not yet evolved warm-bloodedness.

So when did warm-bloodedness first appear? That revolutionary breakthrough must have happened among the successors to the pelycosaurs, the therapsids. We must note as well that the late Pennsylvanian and early Permian, while a time of oxygen high, was a period of temperature low. There was a great glaciation during this interval and much of the polar regions of both hemispheres would have been covered in ice, both continental and sea ice.

The transition from the pelycosaurs to the therapsids is poorly known because of few fossiliferous deposits of the critical age. The gap in our knowledge of the synapsid fossil record extends from perhaps 285 million years ago to around 270 million years ago with some few exceptions in two main regions, the Russian area around the Ural Mountains and the Karoo region of South Africa. The record in the Karoo begins with glacial deposits perhaps as much as 270 million years in age, and then there is a continuous record right into the Jurassic (199 million to 145 million years ago), giving an unparalleled understanding of this lineage of animals.

The therapsids split into two groups, a dominantly carnivorous group and an herbivorous group. By about 260 million years ago the ice was gone in South Africa, but we can assume that the relatively high latitude of this part of the supercontinent Pangea (about 60 degrees South latitude) remained cool. It was still a time of high oxygen, certainly higher than now, but that was changing. As the Permian period progressed, oxygen levels dropped. Seemingly two great radiations of therapsids occurred, among both carnivores and herbivores. From perhaps 270 million to 260 million years ago the dominant land animals were the dinocephalians, and these great bulky beasts reached astounding size, not dinosaur-sized but certainly exceeding the size of any land mammal today save, perhaps, the elephant, and some of the largest of the dinocephalians certainly must have weighed as much as elephants. *Moschops*, for instance, a common genus from South Africa, was 10 feet in length, with an enormous head and front legs longer than the back. This was an enormous animal, the biggest yet on the Earth. But it was graceless in construction, being bulky, and surely awkward. No fast running here, or any deep thoughts, if the tiny size of its brain case is any indication. It was hunted by a group of nearly equally-sized carnivores, also lumbering and slow in all probability.

The dinocephalians and their carnivore predators were hit by a great extinction, still very poorly understood, of some 260 million years ago, the same time, it turns out, that oxygen levels began to plummet. The immediate successors in terrestrial dominance of the dinocephalians, the dicynodonts, were the dominant herbivores of the time from 260 million to 250 million years ago. They in turn were almost eliminated from the planet in the Permian extinction, which will be described in more detail in the next chapter. The dicynodonts were hunted by three groups of carnivores, all therapsids: the gorgonopsians, which died out at the end of the Permian; the slightly more diverse therocephalians; and the cynodonts, which ultimately evolved into mammals during the Triassic. We will return to these groups—and their horrific fate—in more detail in the next chapter.

## OXYGEN AND LAND ANIMALS—REPTILES AND THEIR SIZE

The rise of atmospheric oxygen to unprecedented values of over 30 percent in the Carboniferous-early Permian was accompanied by the evolution of insects of unprecedented size. The giant dragonflies and others of the late Carboniferous through the early Permian were the largest insects in Earth's history. Perhaps it is just coincidence but most specialists agree that the high oxygen would have enabled insects to grow larger, since the insect's respiratory system requires diffusion of oxygen through tubes into the interior of the body and in times of higher oxygen, more of this vital gas could penetrate into ever-larger-bodied insects. So if insects got larger as oxygen raised, what about vertebrates? New data acquired by French anatomist Michel Laurin can be used to test this question. Indeed, it does seem that body size in various late Paleozoic reptilian lineages do track oxygen levels.

Laurin's published data were compared to the oxygen levels given in the Berner curves (like so much discovered while writing this book, this information is simultaneously being published in the scientific literature). Laurin looked at a sample of animals from the anapsids and synapsids. He used a measure of body length and a measure of skull size to evaluate animal size, for specific time horizons between the late Mississippian and the end of the Permian, from about 320 million years ago until about 250 million years ago. Then I did a simple regression analysis of mean size and oxygen levels. Indeed, mean skull size closely tracks oxygen levels, increasing and decreasing in close correspondence with atmospheric oxygen levels for the time periods for which size data is available. As oxygen levels rose in the late Carboniferous, so too did the size of the reptiles increase and, as oxygen began to drop in the mid-Permian, we see size beginning to trend downward. Thus, like insects, it appears that at least these groups of land vertebrates changed size in response to the oxygen available at any given time.

The mammal-like reptiles also seem to show this trend (according to vertebrate paleontologist Christian Sidor of the University of Washington), although the quantitative data (measured in the way that Laurin measured the sizes of earlier land vertebrates) are not yet available. Nevertheless, it is clear that the largest therapsids of all time, the dinocephalians of the middle Permian, evolved at the peak of oxygen abundance. As oxygen began to drop in the mid-Permian, successive taxa assigned to various therapsid groups and, most importantly, the dicynodonts, showed a trend toward smaller skull sizes. While some relatively large forms still lived in the latest Permian—the genus *Dicynodon* and even the carnivorous gorgonopsians come to mind— by this time many of the dicynodonts were smaller. The latest Permian taxa, *Cistecephalus, Diictodon,* and a few others, were very small. The late Permian-early Triassic genus *Lystrosaurus* is smaller in the Triassic than it is in the Permian, and the various cynodonts of the late Permian and early Triassic are all small in size. There are a few giants in the Triassic—*Kannemeyeria* and *Tritylodon* are examples—but in general the therapsids of the Triassic are much smaller than those of the Permian. A recent paper by University of Washington paleontologist Christian Sidor has confirmed the drop in size of Triassic forms compared to their Permian ancestors. Thus, once again, we see there is a strong correlation between terrestrial animal size and oxygen levels, in this case from the latest Permian into the Triassic. In high oxygen tetrapods grew large, and then they grew smaller as oxygen levels diminished.

We have come to a time of about 260 million years ago. It was the end of the long Paleozoic Era, and in the last 10 million years of its nearly 300-million-year-long history all hell would break loose—literally, as the next chapter shall recount.

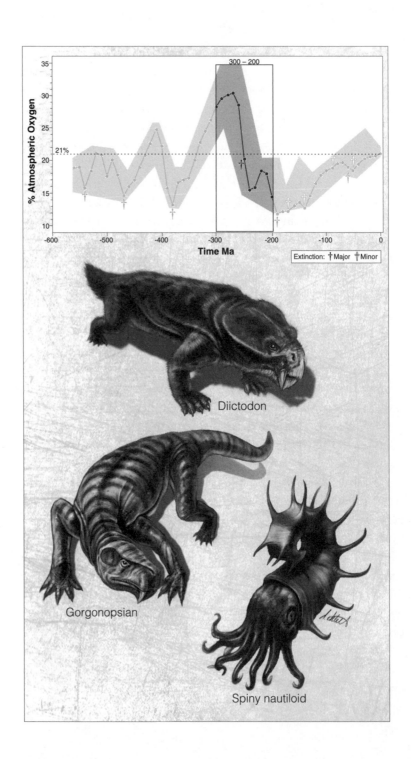

Diictodon

Gorgonopsian

Spiny nautiloid

## 7

# THE PERMIAN EXTINCTION AND THE EVOLUTION OF ENDOTHERMY

A s the Paleozoic came to a close during the Permian Period, the continents all joined into one large landmass and then, some 250 million years ago, an enormous and devastating mass extinction swept land and sea alike. Estimates vary but a figure of 90 percent of all species on the planet is seen as a reliable estimate of the damage done to world species' diversity. The event is the Permian extinction. (It has been given a number of more flamboyant sobriquets: The Great Dying, The Mother of All Extinctions, The Day that Life Almost Died.) Prominent victims in the sea included the majority of brachiopods, all trilobites, all reef-building cnidarians, most crinoids, most bryozoans, many mollusks including the majority of ammonoids, and many fish stocks. On land the extinction was equally devastating, causing great reductions in all stocks of plants and animals. The recovery interval for this extinction was long, and survivors inherited an emptied world.

The cause of the extinction remains one of the most vexing and, to many scientists, a still unsolved geological mystery. Controversy thrives as many different hypotheses compete regarding cause. Not all of them can be correct for some are mutually exclusive. What is not contested is the effect that this paramount extinction event had on the history of life. This chapter looks at both topics—the Permian extinction's real cause and, more importantly, its real effect.

## WHAT HAPPENED?

The Permian extinction came during the greatest relative drop in oxygen (and was also accompanied by a shorter-term rise in atmospheric carbon dioxide, one of the largest in Earth's history). Both of these events took place amid the amalgamation of the largest continental landmass and the greatest volcanic event (the flood basalt eruptions left lava fields the size of Alaska). Are all of these things coincidental? That it was catastrophic is undeniable. It is certainly portrayed as the most catastrophic of the five largest mass extinctions of the past 600 million years by virtually every measure by which extinctions are compared: the percent of the fauna dying out and the effect it had on the nature of the planet's biota.

The study of this mass extinction is ongoing. New work on Permian and Triassic strata in the Karoo of South Africa was instigated by the possibility that meteor impact was indeed the primary or contributing cause. Yet in spite of searching for impact, no evidence for impact has been discovered. Major atmospheric and oceanic oxygen-level changes coupled with global warming remain a hypothesis favored here. We will look at how those twin effects radically changed the nature of life on Earth. First, though, let's go back to a time some 251 million years ago, in the midst of the great Permian extinction.

A voyage 251 million years ago would put us at the very end of the Permian Period. Even at the poles there is no ice. The world is hot and desert-like. There is little plant life, so little, in fact, that soil erosion has caused great dune fields of sand to form. The river systems look like those we saw in our Cambrian voyage where there were no meandering rivers, only ephemeral braided streams—the kind of sheetwash and streamflow that today is found at the bottom of glaciers of alluvial fans—places without vegetation, for it is the roots of plants that allow rivers to have bank stability, which is required for meandering rivers. This place is akin to the time before land plants. Harsh, hot winds, filled with grit, only make the atmosphere seem hotter. And hot it is— a place of high carbon dioxide and low oxygen. And it reeks of rotten eggs. Great bubbles of hydrogen sulfide are periodically emerging from the sea and larger lakes, for we will find that both of these are filled with bacteria producing this deadly gas.

We look closer at the terrestrial world. Where there are copses of vegetation near ponds and water there are herds of the dicynodont *Lystrosaurus*. They are stolid and lamb-sized, even as adults, but nevertheless are the largest animals on the planet. They are not very active, and all are living at sea level. Almost motionless, they slowly graze on the low vegetation; moving about leaves them breathless. Two kinds of small carnivores harass their young—the small cynodonts, looking much like strange primitive dogs, and graceful, low-slung but active diapsids known as *Proterosuchus*. They are ambush predators, for the stalking of prey requires locomotion, and while both are capable of rapid bursts of movement to bring down prey, the low oxygen quickly puts them in respiratory debt. What is striking is that other than these few inhabitants there is little else, and the lack of diversity is perhaps the most striking aspect of this world. A paltry few members of the once common dicynodonts such as *Dicynodon* dodder about, and there is still a last predatory gorgonopsian or two, but these are the last relics of lineages headed to complete extinction. Even insects are rare here and of few varieties, for their kind suffered great losses in the extinction that by this time has been a succession of paroxysms over several million years, and there has still not been an evolutionary burst of new forms. The heat and low oxygen have not relented; in fact, they continue to trend unfavorably. The oxygen content now is even lower than

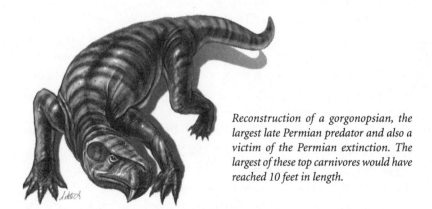

*Reconstruction of a gorgonopsian, the largest late Permian predator and also a victim of the Permian extinction. The largest of these top carnivores would have reached 10 feet in length.*

*Reconstructions of a spiny nautiloid cephalopod, left, and crinoid (echinoderm), right, examples of two groups that nearly ended with the Permian extinction. Only a few species of these and most other animals then on Earth survived to seed the ensuing Triassic period.*

it was 500 million years ago. Things have only gotten worse—and will stay that way for several million years yet.

The sight of the ocean is a shock. It is not blue, but a deep purple in color. The surface regions are cloudy with untold numbers of purple and green bacteria, which reflect the presence of larger quantities of the toxic gas hydrogen sulfide than the low-oxygen water can neutralize through oxidation. Oceans and lakes release the surplus hydrogen sulfide gas and thus poison the land organisms.

We move *into* the sea, and here too the world is reminiscent of the time before animals in many ways. The seabed is covered with widespread stromatolites, the layers of bacteria and trapped sediment that were the main kind of earth life from the period of 3.5 billion to 600 million years ago. The rise of animals in the Cambrian Explosion had seemingly put an end to this formerly common kind of life, since the evolution of grazing invertebrates such as limpets and other snails literally ate this kind of life out of existence. But the herbivorous invertebrates have suffered terrible losses in this Permian extinction, and now the stromatolites have been given new reign once more over the bodies of the newly dead. There are different kinds of survivors among the fish and cephalopods, but there are no coral reefs, or trilobites, or most of the echinoderms that were hallmarks of the Paleozoic. Some brachiopods are left, including large numbers of the very primitive lingulids, and there seem to be a number of small flat clams among them. But like the land, the most striking impression is how few in number the various kinds of life are.

## THE GREATEST OXYGEN CRASH

The two major changes in atmospheric content that seem to have driven the Permian extinction were the rapid drop in oxygen and the rapid rise in carbon dioxide. The interval from about 270 to 200 million years ago was exceedingly interesting. It is evident that carbon dioxide had been as low as it is now for some millions of years prior to the Permian extinction—low enough that it may have greatly impacted the nature of the flora. Our current levels of about 350 to 400 ppm carbon dioxide have been present for perhaps 20 million years, and a major finding of the past decade was the realization that this relatively low level compared to prior times in Earth's history stimulated the evolution of a new photosynthetic pathway. This new kind of photosynthesis, called the C4 pathway, is found in many grasses on our planet. As far as is known, C4 plants did not exist in the Permian and thus the drop of carbon dioxide greatly affected plant life, as shown by the presence of a plant extinction at the time of minimal carbon dioxide levels, at about 305 million years ago. Paleobotanists recognize a mass extinction among plants at this time, with about two-thirds of all plant species known from coal seams going extinct (where, typically, 40-50 distinct plant species can be recovered). While most authors blame this extinction on a drying of the many coal swamps at this time, it seems as likely that the extinction was at least partially caused by the carbon dioxide minimum.

The low levels of carbon dioxide certainly affected global temperatures, precipitating, as we have seen, perhaps the most extensive glaciation of the last 500 million years. The rapid rise of carbon dioxide certainly seems to have coincided with the Permian extinction and brought about rapidly warming conditions on Earth.

The rise in carbon dioxide was certainly dramatic. But it was utterly dwarfed by the fall in oxygen, which may have dropped by two-thirds from its high concentration maximum of as much as 35 percent in the early Permian, to perhaps as low as 12 percent in the early Triassic. The cause of the drop is agreed upon: carbon-rich material was no longer buried at the rates it had been in the late Carboniferous and early Permian. At the same time, the burial rate of pyrite-bearing sediment also dropped dramatically. With huge quantities of reduced carbon exposed to the atmosphere, oxidation set in, removing oxygen molecules from the air. But why did this happen?

The ultimate cause seems to be related to two events. First, the formation of the supercontinent Pangea was completed about the same time that the oxygen drop occurred. As the continents fused together, many of the sedimentary basins and swamps that had been the site of the rapid and profound burial of plant material that caused the rise of oxygen to the maxima of the Carboniferous-early Permian event were uplifted and thus could no longer serve as traps and reservoirs of reduced carbon. Second, the drop in carbon dioxide that culminated about 300 million years ago may have drastically reduced the amount of plant material through mass extinction of species, which in turn caused a significant reduction of plant biomass. There was less plant life to be deposited, and both of these events conspired to cause the oxygen crash.

## POSSIBLE CAUSES

What is called the Permian extinction is really a series of extinctions, beginning at the end of the Guadaloupian Stage of the Permian Period (about 254 million years ago), with a second and far more severe pulse at the Permian-Triassic boundary itself, dated at about 251 million years ago. This "double extinction" at the end of the Permian has been known for about a decade or more. But now we are finding that even between these larger events there were smaller extinction episodes as well. What caused this pattern?

The episode at 251 million years ago is one of the most controversial subjects in modern geological research. There are several scenarios in the controversy, and they are listed below in arbitrary order:

*1.  The mass extinction was caused by a large-body impact with Earth.* Of the many competing hypotheses, the most interesting and controversial is that it was caused by the effects of a large-body impact on Earth some 251 million years ago.

This idea is relatively new. The first paper suggesting this cause appeared only after the turn of the millennium, in 2001. This is the most "journalist friendly" of the various scenarios and is a scientific descendant of the Alvarez Impact Hypothesis of 1980, which had been formulated for the end-Cretaceous catastrophe. The major proponents of impact as the cause of the Permian extinction are Luann Becker of the University of California at Santa Barbara and Robert Poreda of the State University of New York, who in the first part of this century announced that "Bucky Balls" claimed to have been found at Permian-Triassic boundaries at several locales around the globe are evidence of large-body impact some 251 million years ago. About a year after this initial report, which was published in the prestigious journal *Science* and accompanied by worldwide publicity, they announced that they had found the crater as well, a structure named Bedout located near Australia. At that time newly published data on the extinction pattern of marine invertebrates from a stratigraphic section in China indicated that the die-off was sudden and thus consistent with and reminiscent of the pattern of extinction found at the younger Cretaceous-Tertiary boundary sites.

But here the historical parallel between the initial Alvarez discovery and subsequent work markedly diverges with what happened following the Becker et al. announcement. In the case of the Cretaceous work, scientists studying stratigraphic sections at many places around the world found the same evidence that the Alvarez group found: the presence of elevated levels of iridium, the presence of glass spherules and the presence of quartz grains marked by shock lamella. All of these are consistent with impact. But the Becker group's announcement was not followed by paper after paper corroborating their work. No significant iridium was found, no glassy spherules, and a report of shocked quarts (which is consistent with a meteor impact) was later retracted by its author, Greg Retallack of the University of Oregon.

The Bucky Ball story was difficult to corroborate as well, as the protocol that isolates these large carbon lattices is not easily done. The one attempt to replicate Becker's results failed to show any evidence of impact. By 2005 the public and science journalists were the major supporters of the Becker hypothesis, while the working stiff scientists objected—some harshly, which was a response to the tenor of the Becker group and their dogmatic belief in their findings.

2. *Carbon dioxide catastrophe.* This idea, put forth in the late 1990s by Harvard University's Andy Knoll and his colleagues, advocated a rapid and massive release of carbon dioxide that was previously locked up in deep-water Permian sediments. Carbon dioxide poisoning, accompanied by high heat brought about by the greenhouse heating effect of the newly released carbon dioxide, was the proposed kill mechanism. This hypothesis, while attractive, was soon shot down by oceanographers who pointed out that the release of carbon dioxide necessary for this scenario was physically impossible.

3. *Methane catastrophe.* This idea was largely the brainchild of the same Greg Retallack who had mistakenly reported the presence of shocked quartz grains from latest Permian rocks. Retallack noted that carbon isotope values found from most Permian-Triassic boundaries were isotopically so "light" that they could not have been caused by the extinction alone (if all plant life is killed off on a planet, the carbon isotope values go light). But the gas methane is "light," and a massive release of the stuff would produce the observed isotopic signature. Methane is an even better greenhouse gas than carbon dioxide, so Retallack invoked a sudden heating. At the same time, Retallack suggested that the drop of oxygen already thought to have occurred over the time interval of the Permian extinction happened so fast that land animals died of asphyxia. But Bob Berner dismissed this idea of sudden oxygen drop on scientific grounds. And finally, isotope geochemists even began to question the methane idea, recovering many measures that did not yield the methane signature. They ascribed the very light findings to rocks whose carbon isotope values had been perturbed by later heating and pressure.

4. *Heat spike and low oxygen caused by Siberian traps.* With the falsification of the methane idea, attention shifted to the simultaneity of the largest flood basalt of the past 600 million years, the Siberian traps, with the Permian extinction. Because the flow from Earth of

such lava is accompanied by a massive release of carbon dioxide into the atmosphere, it was hypothesized that a sudden spike in global temperatures was somehow involved in the extinction. The oxygen drop was also invoked but not the precipitous drop advocated by Retallack. Instead, the heat spike overlay a long-term drop in oxygen, which began long before the extrusion of the Siberian trap basalts. The extinction pattern under such a scenario would show long-term extinction due to the oxygen drop, followed by (and greatly exacerbated by) a sudden rise in global temperatures. This is the idea that I promoted in two papers published in *Science* in 2005. One showed the extinction pattern of mammal-like reptiles across the Permian boundary in South Africa, the second (published with Ray Huey of the University of Washington) invoked a loss of terrestrial habitat due to the drop in oxygen. As atmospheric oxygen dropped, even moderate altitudes would have had even lower-oxygen content. If oxygen dropped below about 12 percent (the value found by an early version of Berner's GEOCARB modeling), the only place on Earth where terrestrial animals could live would be at sea-level. By using reconstructions of the Permian world, we then showed that sea-level elevations made up about 50 percent of the Permian land surface. Thus, half the land area would be unavailable for animal life, and the areas that were habitable would be cut off from other habitable areas by even modest altitude. The problem with this hypothesis was that while it readily explained the observed pattern of long-term, elevated extinction during the late Permian, the heat spike did not seem severe enough to cause the observed short-term bump in extinction rates at the boundary itself. The effects of a heat spike could be imagined but not modeled.

5. *Hydrogen sulfide poisoning.* The final entry into the Permian extinction sweepstakes was put forward in 2004 by Lee Kump and colleagues at Pennsylvania State University, and is considered here to be the winner of the Permian extinction sweepstakes. Kump and his colleagues suggested that the long period of low oxygen seen both in models and by direct evidence in marine sediments themselves would have created conditions in the sea favoring the massive growth of hydrogen sulfide–releasing bacteria, which overwhelm the surface water's oxygen supply and result in colorful bacterial blooms (hence the sugges-

tion of purple oceans). As anyone who has ever taken chemistry lab knows (or anyone who has had to smell rotten eggs in a closed area can attest), hydrogen sulfide is nasty stuff. Many humans have been killed by high concentrations of hydrogen sulfide found around natural gas wells, especially, coincidentally enough, around the town of Permian, Texas. This hypothesis is thus like the oxygen-drop heat-spike idea above but adds on the pulse of poisonous hydrogen sulfide entering the oceans and atmosphere as the kill mechanism at the boundary itself. As this book is being written, scientists are scrambling to confirm this intriguing proposal as an add-on to cause 4 above. As can be imagined, this new hypothesis is despised by the impact camp but is gaining favor among the rest of us working on the Permian extinction problem. Better than any of the other hypotheses, it explains the relatively sudden death at the Permian boundary, a spike of extinction overlaying a much longer-term interval of heightened species extinction. But we can improve on this.

## A NEW SCENARIO FOR THE PERMIAN EXTINCTION

Here is how the Permian extinction might have occurred. First, it was caused by a succession of similar events, some smaller, some larger. The most damaging (to animals and plants) occurred 251 million years ago, but there were others both before and after, or from about 255 million to 248 million years ago.

Each event began with heat from greenhouse gases rising into the atmosphere. At the same time, the warming ocean began to favor the growth of sulfur-metabolizing bacteria as oxygen levels dropped. If deep-water hydrogen sulfide concentrations increased beyond a critical threshold during oceanic anoxic intervals (times when the ocean bottom and perhaps even its surface regions lose oxygen), then the oceanic conditions (such as those in the modern Black Sea) separating sulfur-rich deep waters from oxygenated surface waters could have risen abruptly to the ocean surface. The horrific result would be great bubbles of highly poisonous hydrogen sulfide gas rising into the atmosphere. The amount of hydrogen sulfide gas entering the late Permian atmosphere would be more than 2,000 times greater than the small modern flux (this is the toxic killer coming from volcanoes). Enough would have entered the atmosphere to most likely lead to toxic levels.

Moreover, the ozone shield, that layer which protects life from dangerous levels of ultraviolet rays, also would have been destroyed. Indeed, there is evidence that this happened at the end of the Permian, for fossil spores from the extinction interval in Greenland sediments show evidence of the mutation expected from extended exposure to high ultraviolet fluxes attendant on the loss of the ozone layer. Today, we see various holes in the atmosphere, and under them, especially in the Antarctic, the biomass of phytoplankton rapidly decreases. If the base of the food chain is destroyed, it will not be long until the organisms higher up are perturbed as well. The complete loss of our ozone layer has even been invoked as a way to cause a mass extinction if Earth was hit by particles from a nearby supernova, which would also destroy the ozone layer.

Finally, an abrupt increase in methane concentrations significantly amplifies greenhouse warming from an associated carbon dioxide buildup and methane levels that would have risen to more than 1,000 parts per million (in fact the carbon dioxide level may have risen to 3,000 parts per million). As the nasty hydrogen sulfide goes into the atmosphere, at the same time destroying the ozone layer, greenhouse gases do their work in making the planet hotter. It turns out that the lethality of hydrogen sulfide increases with temperature, based on hideous lab experiments where various animals and plants are exposed to increasing doses of hydrogen in closed chambers.

## WHY THE PERMIAN EXTINCTION MATTERED

Mass extinctions have long been recognized as potent evolutionary events. Two aspects foster evolutionary change. First, the removal of species through extinction opens the way for new species to form to fill the suddenly emptied niches. The more catastrophic the extinction, the greater this effect will be. The second influence is more gradual. If the mass extinction is caused by a long-term environmental change of some sort, species will have an opportunity to try to adapt to the new conditions. But this effect can only take place given time. An asteroid colliding with Earth gives no warning, and there is no time for any sort of anti-impact adaptations.

Because it was so catastrophic, the Permian extinction certainly affected the biological makeup of the world. The differentiation of a Paleozoic fauna from a Mesozoic fauna allowed John Phillips to both define these terms and subdivide the stratigraphic record into large-scale units—because the change was so striking. The Paleozoic world of trilobites, rugose and tabulate corals, goniatitic ammonoids, and most straight nautiloids, among many, many more kinds of life characteristic of the Paleozoic, were replaced by new, higher taxa.

The changes on land were equally dramatic, with the therapsids being largely replaced by new kinds of reptiles that ultimately gave rise to the dinosaurs, crocodiles, and a host of other Mesozoic icons. All of this change may have been the result of the catastrophic part of the extinction, what could be called the Permian-Triassic boundary event of the longer Permian extinction, the biggest hydrogen sulfide burp of 251 million years ago. But here it is proposed that a second group of evolutionary changes occurred because of the longer-term effects in the late Permian: the changeover from a glacially-gripped world to a hot one—where ultralow carbon dioxide was replaced by high levels of this gas—and perhaps most importantly, it was a world that experienced the most catastrophic drop in oxygen levels known for the past 600 million years, the time of animals. The most consequential change occurring over this interval was the evolution of endothermy, but there may have been equally consequential changes occurring as well, including the evolution of live births. While we have no evidence either way about the latter, there is solid evidence that endothermy appeared in the interval of time when oxygen was dropping fastest. Was this just coincidence, or was it cause and effect? Let's propose a new hypothesis concerning why endothermy was evolved by multiple lines of vertebrate animals.

## METABOLISM AND THE EVOLUTION OF ENDOTHERMY

Here let's return to a subject raised in the last chapter, but one whose history is even more important in this chapter's interval of time than last's. Metabolism is the term used to characterize the acquisition and use of energy by organisms, and metabolic rate is the pace at which acquired energy is utilized. It dictates the amount of fuel consumed and the amount of heat generated. It takes work to conduct the activities of life, which include physical activity, food processing, and tissue synthesis. But even in the absence of these activities, energy is expended in just staying alive—a significant amount of energy, it turns out. The baseline activities of life include ion pumping, protein turnover (they wear out!), blood circulation, and respiration. The minimal level of metabolism that proceeds even in the absence of activity (needed for feeding, defense, and reproduction) and growth is termed basal, or maintenance, metabolism. Much work has gone into figuring out what percentage of total metabolism or energy expenditure is needed for this. Just as a car idling is using up gasoline, so too is just living, with about one-third to one-half of all energy expenditure going to baseline metabolism-burning energy. Living is clearly expensive. Reducing that cost has been a major driver of natural selection: the more efficient an organism is in conserving energy, the more energy there is available for actual activity. Metabolic rate is dictated by only three factors: body temperature, the organism's mass, and the phylogeny (evolutionary history) of the animal in question, and it is of first importance in dictating growth rates and reproductive strategy or pattern.

The highest metabolic rates are observed in animals that maintain a constant body temperature that is independent of ambient, environmental temperature—animals that are described in the last chapter as warm-blooded or, more properly, endothermic. Endothermy is a characteristic of birds and mammals and is considered an advanced trait. Endothermic animals maintain the same internal temperature irregardless of ambient temperature, and thus on cold mornings, when ectoderms are still sluggish from the evening, endoderms are already moving and moving fast—to prey on animals, to avoid predation, to find shelter or mates. There are even more basic advantages of maintaining constant temperature. Metabolism is the work of every cell necessary to stay alive. Each cell needs energy to stay in disequilibrium with the environment, for equilibrium is the state of matter that is not alive. To remain alive, each cell must have a constant supply of energy to run the chemical reactions that are called life. The chemical reactions that keep any cell alive run at different rates at different temperatures, and thus there is a huge advantage to keeping the cell (or cells in a multicellular organism such as an animal) within a fairly narrow chemical range. Thus, endothermy provides the maintenance of relatively constant temperature and thus accomplishes nearly optimal chemical conditions for the various reactions required for growth, reproduction, energy acquisition, and the other aspects of life. But like all "free lunches," there is a hidden cost. Endothermy requires a large metabolic expenditure. Why was endothermy evolved and when?

The fossil record indicates that endothermy evolved near the end of the Permian, as oxygen levels were plummeting. The best indicator of this comes from studies of the early Permian synapsids (such as the fin-backed *Dimetrodon*), which appear to have been ectothermic. So in early Permian times there was no endothermy. But what *is* the first evidence of endothermy? That comes from a small set of bones in the nasal area called turbinals.

The most interesting question that the dicynodonts and cynodonts might be able to answer concerns the advent of endothermy. One means of discerning whether or not a particular group or fossil was warm-blooded comes from the presence or absence of the turbinals. These function to warm air as it passes into the nose in modern mammals (which helps with gas exchange and water retention in the body), and the fossil record shows that even primitive mammal skulls had them. So too did some of the advanced dicynodonts, indicating that endothermy evolved in this lineage. But how early? Perhaps the evolution of endothermy was a consequence of the falling oxygen but rising temperatures of the post-Permian extinction world. Thus, here a new hypothesis can be reiterated, one first mentioned in the already mentioned 2005 *Science* paper by Ray Huey and me. That paper briefly mentioned that turbinals might have evolved because of low-oxygen content. The conventional idea is that they evolved in response to the acquisition of endothermy and helped warm air going in, as mentioned above. But perhaps they evolved to reduce respiratory water loss. During breathing, vertebrate animals experience a loss of body fluid with each exhaled breath. While this is not a huge problem for most animals in our world, in a low-oxygen world animals would have had to breath with more frequency, and this causes greater drying. But an even greater problem would have been the ambient heat. Global temperatures seemed to rise at the same time that oxygen was dropping, causing a deadly combination of high heat and low oxygen. In that context we see turbinals evolving as a water retention adaptation, and only later were they co-opted to help endothermy.

One problem with identifying the presence or absence of turbinals (also known as nasal conchae) is that there were two separate kinds. One type (the simplest) evolved to increase olfactory efficiency by increasing the surface area over which tissue can detect various scent chemicals, while the second kind were used to reduce water loss during respiration. Reptiles show the simplest kinds, which are olfactory. Mammals and birds have the second kind, respiratory conchae, or respiratory turbinals.

The turbinal bones themselves are very fragile and rarely preserved. The ridges they extend from, however, are readily preserved (if sometimes difficult to distinguish from the sediment often filling most skulls). In 1994 Willem Hillenius first showed that true respiratory turbinals existed in late Permian mammal-like reptiles. Two groups of these therapsids showed these: the cynodonts and some therocephalians. Both groups were active predators, and because predation requires more activity than herbivory, it makes sense that these forms, needing more oxygen than herbivores in an oxygen-starved world, would need some way of reducing water loss as they panted their painful way after prey in what would have been like athletics on the tops of the tallest mountains. This finding thus shows that turbinals first occurred when oxygen was fast dropping (and thus is supportive of both views above), that they are an adaptation to low oxygen, and that at this time endothermy was present in more than one lineage and thus independently evolved.

*Hypothesis 7.1:  Turbinal bones evolved in multiple lineages at the end of the Permian as an adaptation for respiratory water retention in a low-oxygen world.*

Based on evidence from the cynodonts and therocephalians, endothermy was present in at least late Permian predatory therapsids and probably in many more groups, and the search for turbinals in various groups around at that time has just begun. Part of the reason for this is that there are so few well-preserved skulls known from the latest Permian rocks. In 1998 I began a research project in South Africa that entailed new collecting of fossil skulls from this interval of time. Over several years we found and collected over 130 skulls. Each skull, however, must undergo a slow and expensive job of preparation, and only some years later do new discoveries become available to therapsid experts. Already this is yielding dividends, though.

Clearly, there is a downside to endothermy—a very high cost in energy that must be expended. So why did it evolve at all? There is an extensive literature dealing with this question and the *what, when,* and *where* questions have been looked at from many different biological angles—save for one. Nowhere is there a discussion of how metabolism and specifically endothermic metabolism would compare to ectothermy in either lower- or higher-oxygen conditions. That will be the slant of this discussion.

In one of the latest written reviews on the origin of endothermy, two experienced workers, Willem Hillenius and John Ruben, have perhaps answered this question but inadvertently. They stated:

> Endothermy is also associated with enhanced stamina and elevated capacity for aerobic metabolism during periods of prolonged activity.

They were addressing the ability of warm-blooded animals to conduct prolonged exercise even as they go anaerobic physiologically. Paradoxically, this characteristic of endothermy also allows an animal to do better at lower-oxygen values. This is evidenced by the fact that larger animals with endothermy can live at higher altitudes (and thus lower-oxygen levels) than equally sized ectotherms. Here we can propose that endothermy evolved in response to the lowering oxygen levels of the late Permian and that the primary advantage of animals with this new adaptation was the ability to remain more active and thus they were competitively superior to the ectotherms of the late Permian.

This can also be viewed in terms of blood flow and heart rate. As anyone who has ever held a small, scared bird well knows, the heart rate of a small endothermic animal can be astonishingly high, well over 100 beats per minute. This allows blood to circulate through the body more quickly, which would be an advantage when oxygen is low. This is part of the reason that birds can live at higher altitudes than similarly sized lizards. In colder temperatures, such as at night, the ectotherms markedly slow their heart rate—and as a consequence reduce the number of molecules of oxygen reaching the cells.

Let's define a hypothesis based on this idea:

*Hypothesis 7.2:  Endothermy evolved in multiple lineages in response to lowering atmospheric oxygen values of the late Permian and came about concomitant with the evolution of a four-chambered heart, body covering, and in some lineages, nasal turbinals bones. The primary reason for this adaptation was not to maintain constant temperature but to increase efficiency in a low-oxygen environment.*

The main proposition for the evolution of endothermy is called the aerobic capacity model, which asserts that endothermy evolved to allow elevated levels of sustainable activity and that the increase in resting metabolic rate shown by endotherms was an accidental by-product. The second possibility proposed in the literature is that endothermy came about to allow thermoregulation. This is very close to the aerobic capacity hypothesis, but it differs in suggesting that it was not so much the ability to increase exercise efficiency (although that would have been important, of course) but the more visceral need to survive in a lowering-oxygen atmosphere. It must be remembered that the lineages of animals found in the late Permian had descended from a time of much higher oxygen than in the late Permian or even today. They came from a time when there was so much oxygen in the air that even very inefficient lungs could easily deliver all the oxygen needed for life. Thus, the drop in oxygen may have been even more calamitous to vertebrates than it might otherwise have been.

The drop in oxygen at the end of the Permian was profound but slow. It was certainly not fast enough to have been a cause of a sudden die-off at the end of the Permian, as advocated by Oregon paleontologist Greg Retallack. But such a slow change is ideal for evolutionary response to changing and deteriorating environmental conditions. Thus, it seems that the Permian oxygen drop, accompanied by the run-up of global temperatures, stimulated various reptile groups to increase their efficiency of oxygen uptake through the evolution of endothermy and by changes in the nasal area of the skull. Soft-part and physiological responses may have taken place as well, such as increasing the number of red blood cells in the body by increasing red-blood-cell-forming bone marrow. The most obvious change would have been in the lungs and circulatory system, and evolution of the four-chambered heart probably happened at this time. But in how many lineages?

## THE FOUR-CHAMBERED HEART

Both the avian and mammalian hearts have four chambers: two auricles and two ventricles. Many other land tetrapod lineages, including many reptiles and all amphibians, use a three-chambered heart. This difference affects the relationship between oxygenated blood coming back to the heart from the lungs and venous blood returning to the heart from elsewhere in the body. The latter is depleted of its oxygen content. In the four-chambered system, there is never mixing of these two blood groups. But in the three-chambered heart, mixing can take place, and this would seem to reduce the efficiency and oxygen-carrying capacity of this kind of respiratory system.

Because the four-chambered heart is associated with endotherms of high metabolic activity, it has long been argued that a perfect separation of bloods was selected for in animals that had higher metabolic activity levels, such as endotherms. Recently, this traditional view has been cleverly questioned by a group of Australian biologists led by Roger Seymour. They have pointed out that some snakes are capable of keeping oxygenated and nonoxygenated bloods separated even though they have a three-chambered heart. Instead, they argue, the four-chambered heart evolved for allowing elevated blood pressure rather than blood itself. Four-chambered hearts are larger than the hearts of ectotherms, and high blood pressure seems to be associated with endothermy.

Again, it must be assumed that this (and all other work about respiration discussed to date) was argued in the context of present-day oxygen levels, since this is never brought up. Seymour and his group pointed out that the activity levels of endotherms require more oxygen getting to various parts of the body faster than in ectotherms. Perhaps it was not so much natural selection for greater activity that drove this respiratory system improvement but the critically low, late Permian oxygen levels. Even with ectothermy, the latest Permian therapsids may have been like human mountain climbers.

## THE ARGUMENTS OVER DINOSAUR ENDOTHERMY

One of the most enduring scientific debates of the past two decades has been about the metabolism of dinosaurs. Were they endotherms, ectotherms, or so massive that neither applied? The arguments have gone back and forth, based on evidence as disparate as bone structure, oxygen isotopes from dinosaur fossils, and reputed predator-prey ratios. Here it is proposed that endothermy originated as an adaptation to low atmospheric oxygen. If that was the case, endothermy should have evolved in multiple lineages near the end of the Permian. We have seen that such evidence exists for the late Permian therapsids, the lineage leading to mammals. But what of the other groups of Permian reptiles, the diapsids and anapsids? There is no evidence one way or another about anapsids, but this is not the case for the other large reptilian group, the diapsids (or archosaurs)—ancestors of crocodiles, dinosaurs, and many other lineages. Until recently most arguments about this lineage have rested on evidence from the modern crocodile group. It is generally agreed that crocodiles are archosaurs belonging to a lineage dating back to the late Permian. According to most phylogenies, this Permian group was also ancestral to the dinosaur-avian lineage and that the fundamental split into separate crocodile and dinosaur-avian lineages took place in the middle to late Triassic. Therefore, late Permian and early Triassic archosaurs were ancestors to both later lineages. So when might endothermy have evolved, if it did at all?

The large number of extant crocodiles are all ectotherms, and because of this it has been theorized that if endothermy evolved anywhere in the archosaur lineages other than in birds it did so only in the dinosaurs. According to this phylogeny, then, endothermy evolved after the crocodile-like lineage split off from the dinosaur-bird lineage. This former group, known as the crurotarsans, evolved into a number of very successful and common taxa of the middle to late Triassic, including the crocodile-like phytosaurs, the wholly terrestrial aetosaurs, and the carnivorous rauisuchians. Endothermy evolved somewhere on the other great branch, known as archosaurs, the lineage leading to dinosaurs and birds. We know that birds are warm-blooded, and in recent decades there has been a great deal of research and speculation as to whether the ancestors of birds, the saurischian dinosaurs, were themselves endothermic. Several camps of dinosaur specialists have formed around this fundamental question about dinosaur metabolism. One group that includes Jack Horner, Robert Bakker, and A. de Riqules argue that dinosaurs were endotherms. More recently, however, a new faction has come forward suggesting that dinosaurs and even the earliest birds were all ectoderms and that endothermy in birds did not arise until at least the Cretaceous Period.

Recently a new hypothesis has been put forward by the same group that argued that late Permian archosaurs had a four-chambered heart and were at least primitively endothermic. This idea has arisen from recognition that, like the contemporaneous therapsids, the late Permian and early Triassic archosaurs had a more upright posture with legs beneath the body, rather than sprawled to the side in lizard-like fashion. The skeletons of both groups suggest an active life style of high mobility. In this model, all the basal archosaurs had warm blood. But later, perhaps in the middle Triassic, the crocodile and crocodile-like lineages returned to a largely aquatic life style, and re-evolved ectothermy, while maintaining the crocodilian four-chambered heart. The argument here is that ectothermy was thus secondarily re-evolved in this lineage for a simple reason: ectothermy aids diving by enabling the animal to stay underwater longer. By reducing oxygen uptake, an aquatic predator can remain underwater longer than can a similarly sized endotherm. Large size also favors diving and breath holding. For every order of magnitude body mass increases, diving time is doubled. Another adaptation to diving is blood "shunting," where oxygenated blood is mixed with less oxygenated blood during dives.

This latter view fits well with the history of the archosaurs. Modern crocodiles have four-chambered hearts, a trait associated with endothermy. Additionally, crocodiles came from ancestors that had an upright rather than sprawling posture. This upright posture is found today only in endotherms.

While the major radiation of the early archosaurs took place in the

*Reconstruction of the common herbivorous therapsid* Diictodon, *a victim of the Permian extinction. The degree of its "mammalness" can only be determined now by characteristics that rarely fossilize, such as hair and internal organs.*

Triassic, they were present in late Permian strata. The oldest Triassic member of the group is *Proterosuchus* from the Karoo of South Africa. Its appearance coincides with the oxygen minimum, and it might be the first of its lineage to have been endothermic.

## RESPIRATORY ADAPTATIONS IN
## LATE PERMIAN-EARLY TRIASSIC THERAPSIDS

As we have seen above, the Permian extinction has been proposed as a time of lowering oxygen. This hypothesis stimulated paleontologist Ken Angelyck of Berkeley to look at various therapsid fossils from the late Permian and lower Triassic to see if there were any anatomical adaptations to lowering oxygen, either long-term or short-term. He examined the skull of every known taxon of therapsid in the latest Permian and early Triassic in various museum collections and came up with a fascinating result. While he could find no short-term changes, the size of the nasal passages and the size of the secondary palate area showed significant increases in size from the late Permian into the lower Triassic. This anatomical change is consistent with and supports the hypothesis of long-term atmospheric oxygen reduction across the Permian-Triassic time interval.

## HABITABLE LAND AREA AT THE END OF THE PERMIAN

A last aspect of the Permian extinction relates to altitude and oxygen. Just as oxygen content diminishes with increasing altitude in our world, so too would a change in altitude during any time in the past act in analogous fashion. But with the rise or drop in oxygen levels, paleoaltitude and its effects on the distribution of organisms would greatly change. Mountain ranges in our world often are barriers to gene exchange, producing different biota on either side of the range. At the end of the Permian just living at sea level would have been equivalent today to breathing at 15,000 feet, a height greater than that found atop Mount Rainer in Washington State. Thus even low altitudes during the Permian would have exacerbated this, so that even a modest set of hills would have isolated all but the most low-oxygen-tolerant animals. The result would be a world composed of numerous endemic centers hugging the sea-level coastlines. The high plateaus of many continents may have been without animal life save for the most altitude tolerant. This goes against expectation based on continental position.

Because the continents 250 million years ago were all merged into one gigantic supercontinent (named Pangea), we would expect a world where there were very few terrestrial biotic provinces, since animals would be able to walk from one side of the continent to the other without an Atlantic Ocean in the way. But altitude became the new barrier to migration, and new studies of various vertebrate faunas appear to show a world of many separate biotic provinces, at least on land. The work of Roger Smith and myself in the Karoo desert, of Mike Benton in Russia, of Christian Sidor in Niger, and of Roger Smith in Madagascar showed that each of these separate localities had distinct and largely nonoverlapping faunas, as predicted by the Huey and Ward model of altitude. This can be formalized as follows:

*Hypothesis 7.3: During times of low oxygen, altitude creates barriers to migration and gene flow. Low-oxygen times there-*

*fore should have many separate biotic provinces, at least on land. The opposite occurs during high-oxygen times: there will be relatively few biotic provinces and a worldwide fauna.*

The drop in oxygen did more than make mountain ranges barriers to migration. It made most areas higher than 3,000 feet uninhabitable during the late Permian-Triassic time interval. Huey and I recognized that there might be a further effect on life's history: it may have contributed to the Permian and Triassic mass extinctions. We called this "altitudinal compression." The removal of habitat because of altitudinal compression would have caused species from highlands to migrate toward sea level or die out. Doing so would have increased competition for space and resources and perhaps would have introduced new predators, parasites, or diseases in the previously populated lowlands, causing some number of species to go extinct. We calculated that by the end of the Permian more than 50 percent of the planet's land surface would no longer have been habitable because of altitudinal compression. There may even have been extinction caused by the effects modeled long ago by Robert MacArthur and E. O. Wilson in their *Theory of Island Biogeography.* These two scientists noted that diversity is related to habitat area and that species died out when islands or reserves of some sort became smaller. Altitudinal compression would accomplish the same by making the continental landmasses functionally lower in terms of usable area.

## THE FATE OF PLANTS

While plants probably were not overtly affected by the gradual drop in oxygen during the latter half of the Permian period, the concomitant rise in temperature coupled with poisonous hydrogen sulfide in the air surely got their attention. Plant species today are highly sensitive to temperature, both high and low, and will migrate to follow their required temperature ranges during climate change intervals. (Unlike mobile animals, which can move into shelter from freezing winds or find shade in blazing heat, plants must just sit there rooted and take it.) The Permian shows two trends—a change in the kind of flora during the period and a substantial extinction of plants at the end.

Tree ferns, seed ferns, and even some of the more archaic plants from the coal ages, such as lycopsids, characterized the cool climate of the early Permian. But as time progressed, conifers, ginkgos, cycads, and other seed plants replaced these floras. This change seems to reflect adaptations to heating and drying, the two climatic trends of the latter part of the Permian. So how severe was the heating? In a study published in 2002, University of Chicago paleobotanist Peter McAllister Rees compiled lists of fossil plant recoveries from around the world during the Permian. This was a Herculean effort of data collections, but its payoff was substantial. Rees showed that during the latter part of the Permian there was a marked shift of high-diversity floras toward higher latitudes—just the sort of pattern that would be expected by a slow but significant global warming. Eventually the tropical latitudes became too hot for most plants, and there would have been giant regions of the globe essentially barren of plant life at the end of the Permian.

This change culminated with the extinction itself. While the death toll varies from place to place, well over half of plant species may have gone extinct, and in the southern hemisphere the total extinction of what has been called the *Glossopteris* flora (glossopterids were a type of woody seed fern that formed forests akin to conifer forests but were lower-growing forms) seemingly all went extinct.

Another curious aspect of the extinction has been the finding of abundant fossils, presumably from fungi at the Permian-Triassic boundary itself, and so global and pervasive is this layer that it has been used for correlation of the boundary. While it was eventually found that this so-called fungal layer was in reality several layers rich in fungal and algae remains, its presence is further evidence that the catastrophic plant extinction that killed plants did not faze them, and with plants gone, the low-growing fungi and algae had no competition for light and nutrients.

## RESULTS OF THE PERMIAN EXTINCTION

While intense controversy still exists about the cause or causes of the Permian extinction, on one aspect of that time interval there is agreement: in the aftermath of the extinction, ecosystems were profoundly affected, and extinction recovery was long delayed. It is this latter evidence that readily distinguishes the Permian extinction from the later Cretaceous-Tertiary event. While both caused more than half of the species on Earth to disappear, the world recovered relatively quickly after the Cretaceous-Tertiary event. As we have seen above, while some earth scientists believe that the Permian and the Cretaceous-Tertiary events were caused by a large-body impact on Earth, it seems as if the environmental conditions causing the Permian extinction persisted for millions of years after the onset of the extinction. It was not until the middle Triassic, some 245 million years ago, that some semblance of recovery seems to have been under way.

These results would be expected if some part of the Permian mass extinction were directly or indirectly caused by the reduction in oxygen at the end of the Permian period. The newest Berner curves show that oxygen stayed low into the Triassic, and there is even some indication that oxygen levels did not bottom out and begin rising until near the end of the lower Triassic, which might account for the long delay in recovery. This evidence suggests that the environmental events that produced the extinction persisted. If so, and if animals were capable of any sort of adaptation in the face of these deleterious conditions, we would predict that the Triassic would show a host of new species not only in response to the many empty ecological niches brought about by the mass extinction but also in response to the longer-term environmental affects of the prolonged extinction event itself. This is the pattern observed for the Triassic—the world was refilled with many species that looked and acted like some of those that were going extinct (therefore an ecological replacement), but there also appeared to be a host of novel creatures, especially on land. The next chapter postulates that many of the new species evolved to counter the continued low-oxygen conditions continuing right into the Jurassic, a period of more than 50 million years. The Triassic was truly the crossroads of animals adapted to two different worlds, one of higher oxygen and one of lower.

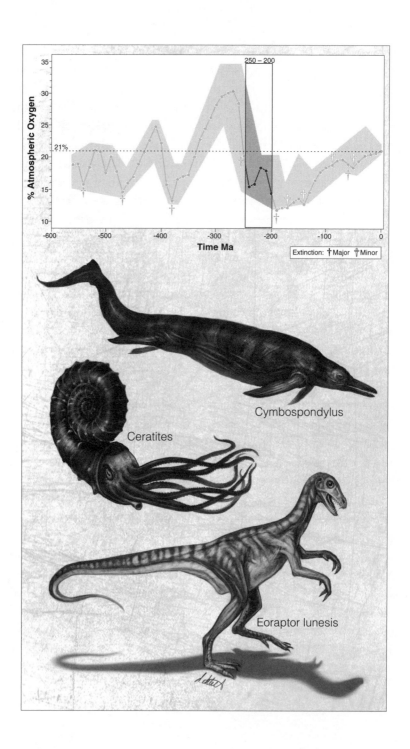

250 – 200

% Atmospheric Oxygen

35
30
25
21%
20
15
10

-600    -500    -400    -300    -200    -100    0

Time Ma

Extinction: † Major  † Minor

Cymbospondylus

Ceratites

Eoraptor lunesis

# 8

# THE TRIASSIC EXPLOSION

The animals and plants of the Triassic Period make up a most interesting assemblage of organisms. This chapter looks at this crossroad in time. Coming out of the most devastating of all mass extinctions, the early Triassic world was empty of life. At the same time, all modeling suggests that a long interval of the Triassic was a time when oxygen levels were lower than today. In Chapter 2 it was proposed that times of low oxygen, especially following mass extinction, foster disparity: the diversity of new body plans. In the Triassic these two factors combined to create the largest number of new body plans seen since the Cambrian. This chapter proposes that it is to that seminal Cambrian time that we can most accurately compare the Triassic. We should give this time and its biotic consequences a new name.

## THE TRIASSIC EXPLOSION

The middle Triassic was a time of amazing disparity on land and in the sea. In the latter, new stocks of bivalved mollusks took the place of the many extinct brachiopods, while a great diversification of ammonoids and nautiloids refilled the oceans with active predators. Fully a quarter of all the ammonites that ever lived have been found in Triassic rocks—a time interval that is only 10 percent of their total time of existence on Earth. The oceans filled with their kind, in shapes and patterns completely new compared to their Paleozoic ancestors—and why not? As shown above, this kind of animal was the preeminent, low-oxygen adaptation among all invertebrates. A new kind of coral, the scleractinians, began to build reefs, and many land reptiles returned to the sea. But it is on land that the most sweeping changes in terms of body plan replacements—and body plan experimentation—took place. Never before and never since has the world seen such a diverse group of different anatomies on land. Some were familiar Permian types: the therapsids that survived the Permian extinction diversified and competed with archosaurs for dominance of the land early in the Triassic, but this ascendance was short lived. Many kinds of reptiles were locked in a competitive struggle with them and with each other for land dominance. From mammal-like reptiles to lizards, earliest mammals to true dinosaurs, the Triassic was a huge experiment in animal design.

Why was this? The conventional answer is that the Permian extinction removed so many of the dominant land animals that it opened the way for more innovation than at any other nonextinction time, perhaps any other mass extinction time as well. It was the most devastating of the mass extinctions. Perhaps, as well, it was simply that many terrestrial animal body plans finally came to an evolutionary point of really working efficiently rather than the sprawling posture of the early reptiles and amphibians. Even as late as the end of the Permian and into the Triassic, groups as mature as the dicynodonts and cynodonts were still trying to attain the most efficient kind of upright posture—rather than the less efficient, splayed-leg orientation of the land reptiles—with all of the ramifications and penalties in respiration that this entailed. But perhaps there is more than this. Body plans were being stimulated into creation by intense selective pressures, and dominant among these was the need to access sufficient oxygen to feed, breed, and compete in a low-oxygen world. There is an old adage about nothing sharpening the mind faster than imminent death. The same might be said about evolutionary forces when faced with the most pressing of all selective pressures—attaining the oxygen necessary for the high levels of animal activity that had been evolutionarily attained in the high-oxygen world of the Permian, when nothing was easier to extract from the atmosphere. The two-thirds drop in atmospheric oxygen certainly lit the fuse to an evolutionary bomb, which exploded in the Triassic.

Thus, the diversity of Triassic animal plans is analogous to the diversity of marine body plans that resulted from the Cambrian Explosion. It also occurred for nearly the same reasons and, as will be shown, was as important for animal life on land as the Cambrian Explosion was for marine animal life. As we saw in Chapter 3, the Cambrian Explosion followed a mass extinction (of the Ediacaran fauna), and it was a time of lower oxygen than today. The latter stimulated much new design. Finally, the Cambrian Period itself ended in a mass extinction—mainly of trilobites that we know of but also among many of the more exotic arthropods known from the Burgess Shale, such as *Anomalocaris*. In similar fashion, on land the Triassic Explosion followed a mass extinction, was a time of lower oxygen, and ended in a mass extinction. Before this extinction, mammals had evolved, as had true dinosaurs, but many of the other kinds of body plans disappeared and dinosaurs, were the dominant land animals. In this end-Triassic mass extinction the dinosaurs suffered least of all. Why dinosaurs? This chapter will look at those questions.

## VOYAGE

Let's begin by looking back to the middle part of the Triassic period. In this middle-late Triassic world, 215 million years ago, on land at least we seem to have arrived among a veritable smorgasbord of animal body plans. Many quite different kinds of vertebrates inhabit this world. Dog-like creatures walk beneath the conifer- and tree-fern dominated vegetation. They are cynodonts, carnivorous varieties, but there are massive herbivores belonging to the same group as well. They are all very mammalian in appearance and behavior, except in one aspect. They move little and seem to tire easily. The carnivores mostly lay in wait, and the herbivores browse stolidly. The cynodonts are not the only mammal-reptiles here, for rhino-sized dicynodonts also browse the low brushy vegetation. Their odd, name-giving tusks extending from a parrot-like beaked mouth make them look like nothing of our world, and seeing them harkens memories of the late Permian world prior to the great Permian mass extinction. All are panting heavily and give the impression of animals having just engaged in strenuous exercise. But most have been motionless; yet they pant for good reason. The level of atmospheric oxygen at this time is equivalent to being higher than 10,000 feet in our world. Except that here we are not atop any mountain: the low swamps and nearby arm of the sea attest to our being at sea level.

Other herbivores are here too, and they are clearly from groups long-diverged away from the mammal-like reptile lineage. It is soon apparent that they are more numerous than the mammal-like reptiles. One of the oddest is another beaked herbivore, from the reptile group known as rhynchosaurs, and near it is a heavily armored quadruped, an aetosaur, looking something like an armadillo, only much larger and better armored. Soon we notice other quadruped reptiles, all fairly primitive archosaurs. Many are 5 to 10 feet in length, and they too move little; when they do, the movement is labored and the panting rapid. Large size and armor evolve for one reason, to avoid being eaten—but the cost is high. Moving a heavy body about extracts a great metabolic cost. Yet there is method in this seemingly morphological madness, for it is clear that the carnivores here are in abundance.

A number of reptiles are visible with heads like that of a crocodile but with bodies obviously evolved for rapid movement on land. Some rise up on their back legs, but they are still quadrupeds for rapid movement. They seem better suited for prolonged movement than the other designs seen till until now, but they are no greyhounds. They prowl but in labored fashion.

Until now all the animals we have seen have been quadrupeds, but it is not long before we encounter our first bipedal animal, and soon we find that this world certainly has its share of animals that walk bipedally. All of them seem far more at ease in this atmosphere, one that bothers even us humans, supposedly advanced mammals that we are. We have finally encountered our first dinosaurs.

Some are small and carnivorous; others, the prosauropods, are relative giants, the largest animals on land and the largest animals ever evolved up to this point. There seem to be many varieties of the smaller bipedal forms, and we are thankful that most of the obvious carnivores are relatively small, for they zip about, running rings around the other vertebrates. But soon a much larger form, 12 feet in length, strides into view, a staurikosaur, the top carnivore among the dinosaurs. We watch this animal, puzzled at how different it seems from the nondinosaurs. Despite its size, it is very active.

This dichotomy extends as well into the swampy and freshwater habitats, where untold numbers of crocodile-like phytosaurs loll on the banks. They too seem not disposed to frolic in any fashion. It is only the dinosaurs that move about with speed, grace, and purpose.

In the oceans we see a major change. Largely gone are the brachiopods, the bivalved invertebrates so dominant in Paleozoic oceans. In their place is another kind of bivalve—the more familiar clams. Few burrow. Most rest on the surface of the sea bottom, sometimes in huge numbers. Swimming above the bottom are many varieties of fish, among them a host of ammonites. This latter group just missed total extinction in the Permian extinction. From the few spared species, however, a host of new species has evolved until now. In the latter parts of the Triassic period, they are even more numerous than anytime before.

There are also coral reefs again, but like the bottom communities of invertebrates found on the sandier and muddier bottoms, the reefs are composed of an entirely new suite of corals. Gone are the tabulate

*Reconstruction of a ceratite ammonite, a group found only during the Triassic. These forms evolved from the few ammonites that survived the Permian extinction, an event that decimated the cephalopods.*

and horn corals, they are replaced by the scleractinians, forms that will persist to our time.

As interesting as the invertebrates and even the fish are of this world, it is the larger vertebrates in the sea that really give pause. Diverse reptilian stocks have obviously returned to the ancestral ocean. Some, the stocky placodonts, resemble large, clumsy seals as they swim down and root through the clam beds with their peg-like teeth. But the stars of the show are the ichthyosaurs, reptiles that have so thoroughly evolved for a swimming habit that they have lost their legs entirely and are the most fish-like of originally terrestrial vertebrates that the world has ever known or will know until the eventual evolution of dolphins, far, far into the future from this Triassic time. While many are small, there are larger forms too: *Mixosaurus* is 30 feet long but it is dwarfed by the monstrous *Shonisaurus*, some of which reach a length of 60 feet. This huge ichthyosaur rivals sperm whales for the title of world's largest aquatic predator of all time, and it preys at will on hosts of fish and smaller reptiles. It has dinner-plate-sized eyes, the largest eyes ever evolved either before or after this.

The marine world, at least, despite its strange and frightful beasts, strikes a sense of some familiarity. But even here we begin to notice strange behavior: the diving reptiles, like the placodonts, come to the surface frequently to breath, as do even the ichthyosaurs. The low oxygen takes its toll.

## TRIASSIC REBOUND

The oxygen story for the Triassic is stunning. Oxygen dropped to minimal levels of between 10 and 15 percent and then stayed there for at least 5 million years, from 245 million to 240 million years ago.

The officially designated early Triassic time interval was from 250 million to about 245 million years ago. During this time there was little in the way of recovery from the Permian extinction. There is also a very curious record of large-scale oscillations in carbon isotopes from this time, indicating that the carbon cycle was being perturbed in what looks like either methane gas entering the oceans, or a succession of small-scale extinctions taking place. All evidence certainly paints a picture of a stark and environmentally challenging world for animal life.

Microbes may have thrived, especially those that fixed sulfur, but animals had a long period of difficult times. But difficult times are what best drive the engines of evolution and innovation and from this trough in oxygen on Earth emerged new kinds of animals, most sporting respiratory systems better able to cope with the extended oxygen crisis. On land two new groups were to emerge from the wreckage: mammals and dinosaurs. The former would become bridesmaids in waiting while the latter would take over the world.

## THE DIFFERENT FATES OF THE
## TRIASSIC THERAPSIDS AND DIAPSIDS

As we saw in Chapter 7, the Permian extinction nearly annihilated land life. The therapsids were hit hard. Much less is known about the diapsids, for at the end of the Permian they were a rare and little seen group in the areas, such as the Karoo or Russia, that have yielded rich deposits with abundant dicynodont (therapsid) faunas. In the Karoo at least, only small fragments of diapsids have come from our uppermost Permian study sections, although two skulls now being prepared as I write this may turn out to be forms that would give rise to the great diapsids and dinosaur dynasty. Roger Smith found them in highest Permian rocks on our last joint collecting trip from the same section where the lowest Triassic diapsids was found. Are they the same species? We will soon know.

If we are still poorly informed about their Permian ancestry, there is no ambiguity about the success of the earliest Triassic diapsids. In the Karoo, in strata only a few meters above the beds that seem to mark the transition from Permian to Triassic, there are relatively common remains of a fairly large reptile known as *Proterosuchus* (also known as *Chasmatosaurus*). This was definitely a land animal with a very impressive set of sharply pointed teeth. It was also definitely a predator, but like a crocodile, its legs were splayed to the sides (if somewhat more upright than the crocodilian condition). But this condition was to rapidly change in the diapsids to a more upright orientation as the Triassic progressed, and more gracile and rapid predators soon replaced the early diapsids such as *Proterosuchus*.

While the need for speed was surely a driver toward this better locomotory posture, just as important may have been the need to be able to breathe while walking. Like a lizard, *Proterosuchus* may still have had a back-and-forth sway to its body as it walked, and this sort of locomotion causes compression on the lung area due to what is known as Carrier's Constraint—the concept that quadrupeds with splayed out legs, such as most (but not all!) lizards, cannot breathe while they run, because their sinuous, side-to-side swaying impinges on the lungs and rib cage, inhibiting inspiration.

For this reason, most lizards and salamanders cannot breathe while walking, and *Proterosuchus* may have had something of this effect, although not as pronounced as in modern-day salamanders or lizards. A solution is to put the legs beneath, but this is only a partial solution. To truly be free of the constraint that breathing puts on posture, extensive modifications to the respiratory system and the locomotory system had to be made. The lineage that led to dinosaurs and birds found an effective and novel adaptation to overcome this breathing problem: bipedalism. By removing the quadruped stance, they were freed of the constraints of motion and lung function. The ancestors of the mammals also made new innovations, including a secondary palate (which allows simultaneous eating and breathing) and a complete upright (but still quadruped) stance. But this was still not satisfactory and a new kind of breathing system was evolved. A powerful set of muscles, known as the diaphragm, allowed a much more forceful system for inspiring and then exhaling air.

Thus, by the middle Triassic, some very different respiratory designs were in play, with natural selection and competition as the arbiters. We know what kind of lungs the mammals had. But what about the diapsids—and their most famous members (and descendents of the early kinds such as *Proterosuchus*)—the dinosaurs? By the end of the middle Triassic they had burst upon the scene. How did they breathe? Therein lies a controversy.

What kind of lungs did the earliest dinosaurs evolve? What changes to this lung design came about in their descendents? This has been the source of controversy for more than two decades now. But before we enter the debate, it should be noted that whichever lung system was found in the Triassic dinosaurs, it evolved for a self-similar reason: oxygen levels reached their lowest point in the Triassic, coincidentally at the time when a majority of Permian land animals were going extinct (leaving many empty niches and thus wholesale evolutionary and ecological opportunity), and many vertebrates responded to these two factors by rapidly producing a host of new kinds of vertebrate body plans—and respiration systems as well. The most famous of the new Triassic body plans was a bipedal form that we call dinosaurs.

## WHAT IS A DINOSAUR?

Because of its general interest and rather sensational aspects, perhaps the most commonly asked question about dinosaurs is the manner of their extinction. The 1980 hypotheses by the Alvarez group that Earth was hit 65 million years ago by an asteroid and that the environmental effects of that asteroid rather suddenly caused the Cretaceous-Tertiary mass extinction in which the dinosaurs were the most prominent victims, keeps this question paramount in people's minds. The fact that this controversy is rekindled every several years by some new finding brings it to the surface once again. Thus its preeminence even supercedes the question of whether or not the dinosaurs were warm-blooded. Way down on the list of questions about dinosaurs is the inverse of the extinction question—not why they died out, but why they evolved in the first place.

We know *when* they first appeared, in the second third of the Triassic Period (some 235 million years ago), and we know what these earliest dinosaurs looked like: most were like smaller versions of the later and iconic *Tyrannosaurus rex* and *Allosaurus*. They were bipedal forms that quickly became large. What has not been largely known or even considered is the new understanding that 230 million years ago was the time when oxygen may have been nearing its lowest level since the Cambrian Period.

So here is a new view here: dinosaurs evolved during, or immediately before, the Triassic oxygen low, a time when oxygen was at its lowest value of the last 500 million years—*and their body plan is a result of adaptation to low oxygen.*

Many other animals changed body plans in response to extremes in oxygen and so too did the dinosaurs, in my view. The dinosaur body plan is radically different from earlier reptilian body plans and appears in virtually a dead heat (and in great global heat) with the oxygen minimum. Perhaps this is a coincidence. But because many of the aspects of "dinosaurness" can be explained in terms of adaptations for life in low oxygen, that seems unlikely.

To formalize this: the initial dinosaur body plan (evolved first by saurischian dinosaurs such as *Staurikosaurus* and the somewhat younger *Herrerasaurus*) was in some part in response to the low-oxygen conditions of the time:

*Hypothesis 8.1: The initial dinosaur body plan of bipedalism evolved as a response to low oxygen in the middle Triassic. With a bipedal stance the first dinosaurs overcame the respiratory limitations imposed by Carrier's Constraint. The Triassic oxygen low thus triggered the origin of dinosaurs through the formation of this new body plan.*

The fossil record shows that the earliest true dinosaurs were bipedal and came from more primitive bipedal thecodonts slightly earlier in the Triassic. These thecodonts (diapsids) were the ancestors of the lineage giving rise to the crocodiles as well and may have been either warm-blooded or heading that way. Bipedalism was a recurring body plan in this group, and there were even bipedal crocodiles early on. Why bipedalism, and how could it have been an adaptation to low oxygen?

Earlier we saw how even most modern-day lizards cannot breath while they run, and this is due to their sprawling gait. Modern-day mammals show a distinct rhythm by synchronizing breathtaking with limb movement. Horses, jackrabbits, and cheetahs (among many other mammals) take one breath per stride. Their limbs are located directly beneath the mass of the body and to allow this the backbone in these quadruped mammals has been enormously stiffened compared to the backbones of the sprawling reptiles. The mammalian backbone bows slightly downward and then straightens out with running, and this slight up-and-down bowing is coordinated with air inspiration and exhalation. But this system did not appear until true mammals appeared in the Triassic. Even the most advanced cynodonts of the Triassic were not yet fully upright and thus would have suffered somewhat when trying to run and breathe.

By running on two legs instead of four, the lungs and rib cage are not affected. Breathing can be disassociated from locomotion. The bipeds can take as many breaths as they need to in a high-speed chase. At a time of low oxygen but high predation, any slight advantage either in chasing down prey, or in running from predators—even in the amount of time looking for food, or how food is looked for, would surely have increased survival. The sprawling predators of the late Permian, such as the fearsome gorgonopsians, were, like most predators during and before their time, ambush predators, as all lizards are today. So what must it have been like for the animals of the Triassic when they found that, for the first time, the predators were out searching for them rather than hiding and waiting? Were any smart enough to register surprise? The results: carnage, surely carnage. This is why dinosaurs may have arisen.

All dinosaurs descended from bipedal ancestors, even the massive quadrupeds of later in the Mesozoic. In the Triassic, the crocodile lineage and the dinosaur lineage shared a quadruped common ancestor. This beast may have been a reptile from South Africa named *Euparkeria*. This group is technically called the Ornithodira, and even the earliest members began to evolve toward bipedalism. This is shown by their ankle bones, which simplified into a simple hinge joint from the more complex system found in quadrupeds. This, accompanied by a lengthening of the hind limbs relative to the forelimbs, is also evidence of this trend, as is the neck, which elongates and forms a slight S shape. These early Ornithodira themselves split into two distinct lineages. One took to the air. These were the pterosaurs, and the late Triassic Ornithodira named *Scleromochlus* might have been the very first of its kind, a still-terrestrial form that looks like a fast runner that perhaps began gliding between long steps using arms with skin flaps. The oldest undoubtedly flying pterosaur was *Eudimorphodon*, also of the late Triassic.

While these ornithodires edged toward flight, their terrestrial sister group headed toward the first dinosaur morphology. The Triassic *Lagosuchus* was a transitional form between a bipedal runner and a quadruped. It probably moved slowly on all fours but reared up on its hind legs for bursts of speed—the bursts necessary to bring down prey, for this was a predator. But it still had forelimbs and hands that had not yet attained the dinosaur type of morphology, so it is not classified as a dinosaur. Its successor, the Triassic *Herrerasaurus*, meets all the requirements and is classified as a dinosaur—the first. As we shall see below, it may have lacked one attribute that its immediate descendents would rectify: a new kind of respiratory system that could handle the still-lowering oxygen content of Earth's atmosphere.

This first dinosaur was fully bipedal, and it could grasp objects with its hands, since it had a thumb like we do. The evolution of this five-fingered hand, being very distinct from the functionally three-toed foot (there were five actual toes, but two were so vestigial that only three toes touched the ground while running or walking), was a new innovation. Because it was not totally bipedal, evolution no longer had to worry about maintaining a hand that had to touch the ground for locomotion. So with a free appendage no longer necessary for locomotion, what to do with it? The much later and more famous *T. rex* reduced the size of the forearm to the point that some have suggested that it was non-functional. Not so for these first dinosaurs, however. While their posture was that of the later carnivorous dinosaurs so familiar to us, their hands were obviously used—probably for catching and holding prey while on the run.

So this is the body plan of the first dinosaurs, from which all the rest evolved: bipedal, elongated neck, grasping hands with a functioning thumb, a large and distinctive pelvis for the massive muscles, and a large surface area needed for those muscles used in walking and running. These early bipeds were relatively small, and before the end of the Triassic they again split into two groups, which remained the most fundamental split of the entire dinosaur clan. A species of these bipedal Triassic dinosaurs modified its hip bones to incorporate a back-turned pubis, compared to the forward-facing pubis of the first dinosaurs. As any schoolboy knows, this change in pelvic structure marks the division of the dinosaurs into the two great divisions: the ancestral saurischians and their derived descendents, with whom they would share the world for about the next 170 million years, the ornithischians.

Of interest here, of course, is how dinosaurs breathed. First, we must look at vertebrate lungs and then tackle the controversy over dinosaur lungs.

## AMNIOTE LUNGS

The lungs of modern-day amniotes (reptiles, birds, and mammals) are of two basic types (although we will see that there are more than two respiratory systems, which include lungs, circulatory system, and blood pigment type). Both kinds of lungs can be reasonably derived from a single kind of Carboniferous reptilian ancestor that had simple sac-like lungs. Extant mammals all have *alveolar* lungs, whereas extant turtles, lizards, birds, and crocodiles—all the rest—have *septate* lungs.

*Alveolar Lungs* These lungs consist of millions of highly vascularized, spherical sacs called alveoli. Air flows in and out of the sacs; it is therefore bidirectional, a characteristic of alveolar lungs. Mammals use this system, and our familiar breathing in, out, in, out is quite typical. Nothing extraordinary here about us humans (compared to most other mammals, that is). The trick, of course, is that air must be pulled into these sacs and then expelled again as oxygen switches place with carbon dioxide. We do this by a combination of rib cage expansion (powered by muscles, of course) and contraction of the large suite of muscles collectively called the diaphragm. Somewhat paradoxically, contraction of the diaphragm causes the volume of the lungs to increase. These two activities, the interacting rib expansion and diaphragm contraction create a reduction in air pressure in the lung volume and air flows in. Exhaling is partially accomplished by elastic rebound of the individual alveoli. When they inflate, they enlarge, and soon after they naturally contract due to the elastic properties of their tissue. The many alveoli used in this kind of lung allow for a very efficient oxygen acquisition system, which we warm-blooded mammals very much need in order to maintain our active, movement-rich life styles.

*Septate Lungs* In contrast to the mammalian lung, the septate lung found in reptiles and birds is like one giant alveolus. To break it into smaller pockets that increase surface area for respiratory exchange, a large number of blade-like sheets of tissue extend into the sac. These partitioning elements are the septa, which give these kinds of lungs their name. There are many variations on this basic lung design among the many different kinds of animals that use it. Some kinds of septate lungs are partitioned into small chambers; others have secondary sacs that rest outside the lung but are connected to it by tubes. As in the alveolar lung, airflow is bidirectional but a difference is that the septate lungs are not elastic and thus do not naturally contract in size following inhalation. Lung ventilation also varies across groups with the septate lung. Lizards and snakes use rib movement to draw air in, but as we have seen, locomotion in lizards inhibits complete expansion of the lung cavity, and thus lizards do not breathe while moving.

The variety of modifications of the septate lung makes this system more diverse than the alveolar system. For instance, crocodiles have both a septate lung and a diaphragm—an organ not found in snakes, lizards, or birds. But the crocodile diaphragm is also somewhat different from that in mammals. It is not muscular but is attached to the liver and movement of this liver/diaphragm acts like a piston to inflate the lungs, with muscles attaching to the pelvis. (The mammalian [including human] diaphragm pulls the liver in the same way a crocodilian one does, creating a visceral piston, but the way this is accomplished differs in crocodiles and mammals.)

*The Avian Air Sac System* The last kind of lung found in terrestrial vertebrates is a variant on the septate lung. The best example of this kind of lung, and its associated respiratory system, is found in all birds. In this system the lungs themselves are small and somewhat rigid. Thus, bird lungs do not greatly expand and contract as ours do on each breath. But the rib cage is very much involved in respiration and especially those ribs closest to the pelvic region are very mobile in their connection to the bottom of the sternum, and this mobility is quite important in allowing respiration. But these are not the biggest differences. Very much unlike extant reptiles and mammals, these lungs have appendages added known as air sacs, and the resultant system of respiration is highly efficient. Here is why.

We mammals (and all other nonavians) bring air into our dead-end lungs and then exhale it. Birds have a very different system. When a bird inspires air, it goes first into the series of air sacs. It then passes into the lung tissue proper, but in so doing the air passes but one way over the lung, since it is not coming down a trachea but from the attached air sacs. Exhaled air then passes out of the lungs. The one-way flow of air across the lung membranes allows a countercurrent system to be set up—the air passes one direction, and blood in the blood vessels in the lungs passes in the opposite direction. This countercurrent exchange allows for more efficient oxygen extraction and carbon dioxide venting than are possible in dead-end lungs. The air sacs are not involved in removing oxygen; they are an adaptation that allows the countercurrent system to work.

There is no question that the greater efficiency of this system (compared to all other lungs in vertebrates) is related to the two-cycle, countercurrent system produced by the air sac-lung anatomy in birds. But when did this system first appear and in how many groups? Therein lies the controversy.

The bones that in birds house air sacs (or parts thereof) should provide fossil evidence. But does a hole in a bone mean there were air sacs? Some of the air sacs, such as the abdominal air sacs in modern birds, leave no record of themselves in hard parts and thus could have been present in extinct groups without leaving a trace. So what was the situation in dinosaurs? There are two different camps: one proposes that dinosaurs had air sacs (Bakker and Paul, among others), the other that the dinosaurs had a simpler reptilian lung system (John Ruben and others).

When did this superb adaptation for low oxygen first evolve? This question is at the crux of a very contentious dispute between three scientists who have taken two very different positions about the kind of lungs that dinosaurs may have had. On one side is John Ruben of Oregon State University, an expert on reptile physiology and respiration, who contends that the flow through avian lung with its many auxiliary air sacs did not appear until the Cretaceous, some 100 million years ago—and then was found only in birds of that time. On the other side of this dispute are two workers profoundly interested in dinosaurs—Robert Bakker, author of *The Dinosaur Heresies* and champion of warm-blooded dinosaurs, and Gregory S. Paul, noted artist and student of predatory dinosaurs, whose book *Dinosaurs of the Air* explicitly looks at this problem.

There is no love lost between these two sides, and the controversy has spilled over into the many Internet dinosaur posts. But an interesting new angle to this debate is that neither side entered into the argument by considering that oxygen levels may have fluctuated in the past, or even that the different kinds of lungs would favor or inhibit various kinds of animals living at different altitudes. In fact, it appears that both sides implicitly believe that oxygen levels in the early to middle Mesozoic may have been *higher* than now. This topic is very pertinent to our discussion here.

## AIR SACS IN DINOSAURS?

As we saw above, the respiratory system of modern birds is composed of small lungs that have appendages—air sacs, which are also used for respiration. Together, the lungs and air sacs extract more oxygen than do the lungs of any other land animal. It has been estimated that at sea level a bird is 33 percent more efficient at extracting oxygen from air than a mammal. At higher altitudes the differential increases. *At 5,000 feet, a bird may be 200 percent more efficient at extracting oxygen than a mammal.* This gives birds a huge advantage over mammals at living at altitude. If such a system were present in the deep past, when oxygen levels even at sea level were lower than are found today at 5,000 feet, surely such a design would have been advantageous, perhaps enormously so, to a group that had it in competing or preying on groups that did not.

We know that birds evolved from small bipedal dinosaurs that were of the same lineage as the earliest dinosaurs—a group called saurischians. The first bird skeletons are from the Jurassic. But the air sacs attached to bird lungs are soft tissue and would fossilize only under the most unusual circumstances of preservation. Thus, we do not have direct evidence for when the air sac system came about. But we do have indirect evidence, enough to have stimulated the air sac in dinosaurs group to posit that *all* saurischian dinosaurs had the same air sac system, as do modern birds. And, like birds, they were warm-blooded. The evidence comes from holes in bones, places where these air sacs may have rested. The Ruben group, however, vehemently opposes this view, suggesting, instead, that the air sac evolved only in the Cretaceous. A very distinct line in the sand was drawn, and the resulting controversy became juicy and nasty and, as will be shown below, is now resolved. Let's look at this controversy in detail, as it impinges on many of the themes of this book so far.

The avian respiratory system with its various air sacs has been called an air sac system, and subsequently we will refer to it as such. The debate about when it first evolved began in the early 1970s with a young and energetic Robert Bakker, who made the breathtaking suggestion that some dinosaurs had the air sac system too. His evidence was as follows.

It had been known since the late 1800s that some dinosaur bones had curious hollows in them—just as bird bones do. For decades this discovery was either forgotten or attributed to an adaptation for lightening the massive bones, for many of these bones with holes—later called pneumatic bones—came from the largest land animals of all time, the giant sauropods of the Jurassic and Cretaceous. The pneumatic bones were found mainly in vertebrae. Birds have similar pneumatic vertebrae, and while it can be said that some bird bones were light to enhance flying, it was also clear that some of the air sacs attached to bird lungs rested in hollows in bones. Thus, in birds, bone pneumaticity was an adaptation for stashing away the otherwise space-taking air sacs. The bodies of animals are filled with necessary organs and putting the air sacs in hollowed-out bones made a lot of evolutionary sense. But Bakker made the leap and suggested that the pneumatic bones in his beloved fossil sauropods had evolved for a similar purpose and were direct evidence that sauropods had and used the air sac system. He was quite explicit, writing (in his stimulating 1986 book, *The Dinosaur Heresies*):

> The dinosaur's vertebral hollows are so similar to birds' that there can be little doubt an avian style system of air sacs was at work in these Mesozoic animals. Moreover, the holes in bones represent the periphery of the total system. Birds locate their largest air sacs between their flight muscles and in their body cavity. Some dinosaurs (duckbills, horned dinosaurs) exhibited no vertebrate body but I suspect they had located air sacs fully within their body cavity.

Bakker thus put not only those dinosaurs with pneumatized bones (the group known as saurischians) into the air sac contingent but also the other major dinosaur group, the ornithischians, as well.

Bakker's larger purpose was to add further evidence that dinosaurs were warm-blooded. Birds, with their enormous energy and oxygen demands related to flying, were thought to have evolved the air sac system as a way to satisfy the metabolic demands of their endothermy. Thus, if Bakker could show that dinosaurs had this respiratory system too, his arguments about the presence of endothermy in dinosaurs carried more weight. He even makes a comment about "richly oxygenated" Mesozoic air, supposing (when he wrote this in 1986) that oxygen levels were higher than present-day levels.

Following Bakker, other dinosaur workers took up the call and the specific case of air sacs being present in sauropods was most recently (and most forcefully) espoused by Berkeley dinosaur paleontologist Matt Wedel in 2003. But by far the most eloquent and substantial arguments for the air sac system being present in the dinosaurian precursors to birds came from Gregory Paul, who in 2002 published his massive and lavishly illustrated *Dinosaurs of the Air: The Evolution and Loss of Flight in Dinosaurs and Birds.* In 460 pages, Paul looked at every aspect of the dinosaur to bird transition and the evolution of the air sac system in saurischian dinosaurs, long before the evolution of the first bird, was dealt with in many pages of intricate detail. We need to look at his arguments.

Paul makes many points, several are central to the argument. First and foremost, he points out that the air sac system in flying birds must show adaptations to allow (or at least not impede) flight. He also noted that nonflying birds have air sac systems anatomically distinct from flyers. Thus, the bones of dinosaurs would not necessarily show all the adaptations for air sacs seen in flying birds—that attainment of flight itself would require adaptations to the air sac system not required by nonflyers. Here is the list of characteristics that he finds associated with the air sac system:

- Pneumatic bones, especially in the vertebral column.
- Shortening of the trunk of the body.
- Shortening of the first dorsal ribs.

- Elongation and increased mobility of posterior ribs. This mobility is enabled by the presence of ribs with double heads at their ends.
- Uncinate processes on several of the ribs (these are small, hook-shaped bones attached to the trunk ribs).
- A hinge joint making up the attachment of the ribs with the sternum.

## EVOLUTION OF THE AIR SAC SYSTEM

The complex air sac-lung system found in birds had to have evolved from a reptilian, sac-like lung. Here is the pathway envisioned by Paul in his 2002 book.

The first of the so-called archosaurs were the primitive late Permian through early Triassic reptilian group (that we have called diapsids), which would eventually give rise to crocodiles, dinosaurs, and birds. Examples of this group included the quadruped form *Proterosuchus* (described above as one of the earliest Triassic archosaurs). They would have had a reptilian septate lung. Inspiration may have been aided by a primitive abdominal pump-diaphragm system (more primitive, perhaps, than the system found today in modern crocodiles). Successively, however, evolution of the air sac system may have progressed fairly rapidly. By the time the first true dinosaur was seen in the middle Triassic part of the air sac system may have been in place.

The most primitive theropods from this time (the first dinosaurs) do not show bone pneumatization, but Paul suggests that the lung itself may have become inflexible and smaller—a bird-like trait based on rib anatomy. The ribs also become double headed, showing that the rib cage itself was capable of a great ventilation capacity. Perhaps as a consequence of going bipedal, these dinosaurs may have switched from the more primitive abdominal pump system to the first air sac system—one with only the abdominal air sac found in modern-day birds. Soon after, descendents of these first dinosaurs, forms such as the well-known, upper Triassic *Coelophysis*, showed the evidence of bone pneumatization, consistent with the proposal that more air sacs had evolved, this time in the neck region. With the Jurassic forms such as *Allosaurus*, the air sac system may have been essentially complete (but still much different from the bird system, modified as it has been for flying, for even the modern-day flightless birds came from flyers in the deep past), with large thoracic and abdominal air sacs. Yet holes in bones do not an air sac system make, if I may paraphrase Yoda.

By the time *Archaeopteryx* had evolved in the middle part of the Jurassic, there may have been a great diversity of respiratory types among dinosaurs, some with pneumatized bones, some without. There also may have been a great deal of convergent evolution going on. For instance, the extensive pneumatization in the large sauropods studied with such care by Wedel may have arisen somewhat independently from the system found in the bipedal Saurischians.

Paul considers the evidence at hand as proof of a progressively more complex air sac system appearing in middle Triassic to Jurassic dinosaur lineages. He summarized this view in his 2002 book:

> One could hardly ask for a better pattern of incremental evolution progressing to the avian skeletal features needed to operate respiratory air sacs. This fact reinforces the case for pre-avian pulmonary air sac ventilation in predatory dinosaurs. No evidence for progressive evolution of a pelvis based diaphragmatic muscle pump (the system found in modern crocodiles) has been presented.

Yet for all these arguments what the paleontologists had were a series of holes in bones, for in no case was a fossil air sac to be found (nor was one expected to be found, of course). And as might be expected, it was not long before a spirited opposition sprang into action. The leader of that opposition was the already introduced John Ruben. During the 1990s he began an extended debate with the advocates of a pre-avian air sac system in general and Greg Paul in particular.

John Ruben and various coauthors came to an opposite conclusion about almost every aspect of what might be called the "air sac in dinosaurs" hypothesis. And they went well beyond even that. In a summary paper published in mid-2005, Ruben and three coauthors proposed that dinosaurs were ectothermic, as were the earliest birds. According to this idea, birds gained warm-bloodedness and the air sac system only with the evolution of flight, and thus warm-blooded, bird-lunged (air sac) birds may not have evolved until the Cretaceous Period, many millions of years after evolution of the first birds. Ruben and his colleagues proposed that dinosaurs possessed simple, septate lungs that were ventilated with the same system now found in crocodiles—a hepatic air pump diaphragm, that operates by muscles attached to the pelvis. While admitting that some dinosaurs had pneumatized bones, Ruben does not think this is evidence of air sacs. But if not air sacs, what system was used by the dinosaurs? According to Ruben and his colleagues, we have only to look at respiration in crocodiles to see how—and with what organs—dinosaurs breathed.

We have already described the crocodilian system, called the hepatic (liver) piston pump. Like mammals, a diaphragm system inflates the lungs, but, unlike our system, the crocodiles move their entire liver region, as their diaphragm muscles attach partly to the pelvis and partly to the soft tissue of the liver itself along a broad band. The entire liver is pulled back toward the pelvis like a piston, and in so doing the lungs inflate with air. But a piston must fill its cylinder, and such a system would by necessity have a partition right across the body cavity—essentially a subdivision of the forward or thoracic part of the interior of the body from the posterior or abdominal region. Crocodiles have this, and Ruben et al. think that dinosaurs breathed in the same way. They also cite aspects of dinosaur skeletons that seem to preclude the possibility of abdominal air sacs at least. As we have seen, the ribs of birds are capable of extensive movement because of a hinged contact between the posterior ribs and the sternum (the breastbone).

As a further nod to the crocodiles, the Ruben camp does not support warm-bloodedness in dinosaurs. Thus, their views could hardly be more different from the air sac dinosaur endothermy advocates. *Voila*—there could hardly have been a better recipe for controversy. Unfortunately, neither camp could land any knockout blow to the other's prime hypothesis based on the evidence at hand. And that is why the discovery of exquisitely preserved feathered dinosaurs from China brought the controversy to a boil.

## ENTER THE CHINESE DINOSAURS

The late-twentieth-century discovery of exquisitely preserved bipedal dinosaurs from a spate of quarries in China of early Cretaceous age seemed like exactly the evidence needed to conclude the air sac debate. A requirement of the hepatic piston system, as we saw in the previous paragraph, is a separation of the body cavity into forward and rear parts by some sort of partition. One of these newly collected specimens, named *Sinosauropteryx,* showed this type of partitioning—according to Ruben et al. in 1997 and repeated in a review published in *Science* in 2005. They saw a dark carbonized region in a position within the body cavity where just such a partition is found in crocodiles and because of this they argued that the case was closed—and crowned themselves victors. Paul, of course, had a different view. He looked at this specimen and noted that it had been broken during its collection into many individual pieces and that the required reconstruction of the fossil may have produced the critical black region and thus (in his opinion) could not be cited as evidence of the hepatic piston system. Paul then cited an even more elegantly preserved specimen of a small dinosaur known as *Scipionyx,* which, according to him, shows the presence of a fossilized air sac!

So back and forth, statement and rebuttal, at this writing the battle rages still (although one side does not seem to know that it has lost). But as noted above, none of the participants considered the effect that potentially low oxygen may have had on the evolution of the early dinosaurs. In fact, Paul, like Bakker before him, cited an older (and now discredited) reference to the Mesozoic having had *higher* oxygen than now, based on findings of air bubbles trapped in amber. But results of the GEOCARBSULF work indicating oxygen levels during these evolutionary events tell a very different story. It may have been that this prolonged period of low oxygen, more than any other factor, caused what can be called the Triassic Explosion—one part being the evolution of dinosaurs. That said, could the dinosaur air sac question be resolved? Ruben is correct in stating that the presence of pneumatic bones does not "prove" the existence of air sacs in dinosaurs. (Very few scientists ever prove anything.) But the progressive increase in pneumatization in the saurischians strongly supports the theory of air sacs in dinosaurs.

## AIR SAC CONTROVERSY RESOLVED?

A really good paper in *Nature* or *Science* can tip the scales of any controversy, because publication of any controversial topic in either of these journals occurs only after rigorous review. A paper in *Nature* from July 2005, titled "Basic avian pulmonary design and flow-through ventilation in non-avian theropod dinosaurs," by paleobiologists Patrick O'Connor and Leon Claessens brought new morphological data to the controversy and must have felt like a stake in the heart to dear John Ruben.

The paper did not stake out new territory. It looked at existing lines of evidence in a new way, using a nearly unbelievable amount of data collection to support the contention that saurischian dinosaurs (all except the very first, that is) most probably had the avian air sac system. Here is what O'Connor and Claessens did. They looked at birds—a lot of different birds, more than 300, in fact. They looked at birds in more detail than anyone ever has. And by "looking," the two scientists poked, prodded, filmed, dissected, and injected vast quantities of gooey plastic into the respiratory systems of so many different birds that at the end of the paper they had to plead that all the poor bird "subjects" were dead or dying anyway (of course, they were dead or dying after having their air sacs filled with rubber). But the most amazing thing emerged. The avian air sacs are *way* more voluminous and complicated than anyone had suspected. By injecting the air sacs with latex, each tiny evagination of an air sac into the bird's skeleton could be wonderfully observed and characterized. For perhaps the first time, the real relationship of air sac to bone—the pneumatization that we have talked about for so many pages now—could be observed. They even attached a movie of bird respiration, as observed with radiography, in the online version of their paper—thus the birds. But the crux of the paper was in showing that the very specific shapes required for placing parts of the air sacs in bird bones are found in saurischian dinosaur bones. Not just holes in dinosaur bones, but *the same shapes of holes in the same (or homologous) bones.*

The folks arguing that there was no air sac system in dinosaurs have not denied that dinosaur bones had holes in them. They said the holes were there all right but that they were adaptations simply for lightening the bones. But there comes a point when an argument finally collapses under the weight of too great a coincidence. Like a bad movie in which the plot depends on some coincidental happenstance far too outrageous for ordinary life, here we would have to accept the coincidence that the same shapes used for lightening bones are also optimal for storing bits of respiratory structure.

The final bit of thunder in O'Connor and Claessen's paper was in its description of the holes found in an exquisitely preserved early saursichian dinosaur. Because of its importance in the most crucial debate of these middle chapters we can quote from the two authors:

> Here we report, on the basis of a comparative analysis of region-specific pneumaticity with extant birds, evidence for cervical and abdominal air-sac systems in non-avian theropods (saurischians), along with thoracic skeletal prerequisites of an avian-style aspiration pump (the air sac respiratory system). The early acquisition of this system among theropods (bipedal saurischians, the first dinosaurs to evolve) is demonstrated by examination of an exceptional new specimen of *Majungatholus atopus*, (a primitive saurischian dinosaur) documenting these features in a taxon only distantly related to birds. Taken together, these specializations imply the existence of the basic avian pulmonary Bauplan (body plan) in basal neotheropods, indicating that flow-through ventilation of the lung is not restricted to birds but is probably a general theropod characteristic.

A blow had been struck for dinosaurs (or saurischians at least) having the avian, air sac respiratory system. But was it a knockout punch, the kind of data that ends a debate? Or was this the start of another inconclusive round of partisan clinching? The new study made clear that the Ruben group had made a crucial error in asserting that the bone holes in birds only contained the cervical or neck air sacs. For instance, in an influential 2003 paper, Ruben and coauthors Terry Jones and Nicholas Geist wrote:

> Pneumatization of the avian skeleton, with the exception of the long bones of the hindlimbs in a small subset of birds, is limited to the axial skeleton and forelimbs and results from invasion *by the anterior (cervical and clavicular) air sacs but is not linked to respiratory function or specific lung morphology.* (I have added the italics.)

"Wrong!" argued O'Connor and Claessens. Their exacting observations on living birds showed that holes (pneumatization) in the bones of the middle part of the bird spine were caused by invasion of a *different* set of air sacs than those suggested by Ruben et al.—most crucially sacs found in the abdominal regions—and which are indeed importantly linked to respiratory function and the very specific lung morphology. And the bipedal saurischians showed a virtually identical set of holes in the same bones.

What about the ribs in the saurischians? Would they have the mobility necessary to allow inflation of the air sacs? This too seems to be answered by the two authors. They found that the posterior ribs showed a change in attitude that made the air sac system permissible. Also potentially involved are the "gastralia," free-floating bones found in some dinosaurs.

Perhaps the greatest contribution of this paper was in pointing out that the entire community has misunderstood which air sacs penetrate which bones in birds. Now it is understood that it is the sacral, or tailward, parts of the body that are most important in producing the characteristic "bird-breathing" pattern. So it seems that the case is closed—at least for saurischians. For the ornithischian dinosaurs there is little evidence for the air sac system, and as we shall see, this meshes well with their distribution in time. During the Jurassic times of very low oxygen they were minor elements of the fauna. It was not until the great oxygen rise of the late Jurassic through the Cretaceous that this second great group of dinosaurs became common.

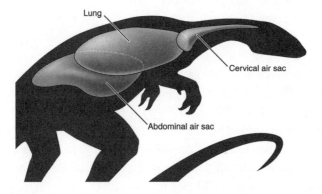

*Air sac system in a saurischian dinosaur, as envisioned by O'Connor and Claessens in their 2005* Nature *paper. The cervical (neck) air sacs are present. But also shown are the mid-body and abdominal systems, previously only surmised to be in these kinds of dinosaurs. It appears that the saurischians, like modern birds, were superbly adapted for low oxygen, and this probably accounts for their success.*

## TRIASSIC- MIDDLE JURASSIC DINOSAURS: A NEW KIND OF ANIMAL WITHOUT MODERN COUNTERPART

It is time to come to the heart of the matter about dinosaurs—or at least the earliest forms. The most important questions are whether or not they were warm-blooded and what kind of respiratory system they had in terms of lung morphology, circulatory system morphology (especially the heart), and physiological adaptations for low oxygen. First, let's look at the environmental context of their first evolution and existence.

The Triassic was a time of low oxygen—which we can find today at high elevations—but of atmospheric gas pressure equal to, or perhaps even exceeding, that of today. This is beyond our experience. It was a time of very high carbon dioxide. Our species has never experienced this (yet we are trying to make it so). It was also a time when the climate of the planet was hot—much hotter than today: no icecaps, no glaciers even in the mountains, and heat from equator to pole. Is it any surprise, then, that the animals of that time may have been unlike anything of the modern-day?

Perhaps the best reference for all things dinosaurian comes from the marvelous, weighty tome *The Dinosauria* (Second Edition). Edited by David Weischampel, Peter Dodson, and Halszka Omolska, this compendium has data that are currently being mined to show many trends, and some of these data have already been used in the pages above and the pages to come. But in addition to raw data on dinosaur ranges and geographic occurrences, there are summary chapters by leading specialists on all of the issues invoked above, the only difference being that none of the authors takes into account the radically low-oxygen levels during the time of the dinosaurs' first evolution, conditions that carried on for many tens of millions of years of the Mesozoic. On the twin and central issues of metabolic rate (endothermy versus ectothermy) and respiratory adaptations, there are assured stances that we should look at.

Phylogenetically, dinosaurs are positioned between crocodiles and birds, the former cold-blooded with hepatic piston lungs, and birds with warm blood and the air sac system. It is thus fair to ask where dinosaurs sit in terms of these two end members: Were they closer to crocodiles or closer to birds based on morphological evidence? And which of the adaptations—cold blood with relatively simple tidal lungs or warm blood with radically transformed and efficient (for low oxygen) lungs did the first dinosaurs possess?

With regard to metabolism, the chapter by dinosaur experts Kevin Padian and Jack Horner points out that dinosaurs were probably more than "damned good reptiles," citing the rapid growth rates observed in dinosaurs. But were they warm-blooded? Birds are, and it is inferred by Padian and Horner that the direct ancestors of birds, the nonavian bipedal saurischians of the middle Jurassic, may have already evolved this characteristic. But by the time of the first birds (*Archaeopteryx*) of the late Jurassic, oxygen levels had dramatically risen from their late Triassic nadir. The world was a very different place from the Triassic world, it was one closer to our own, where oxygen was no longer a limiting factor for metabolism.

As far as lungs and respiration go, the chapter by Anusuya Chinsamy and Willem Hillenius took a supposedly dispassionate look at the possibility of an air sac system in dinosaurs that ended up sounding very much as if it had been written by John Ruben. They took the conservative route—no air sacs. And what about the possible presence of air sacs? We know that air sacs ultimately evolved in birds and that the presence of bone pneumaticity makes a strong case for the air sac system. A fully evolved air sac system would have been the best way to deal with the low-oxygen conditions at the time of the dinosaurs' first appearance and early history, from the late Triassic to the upper Jurassic. But even the true believers, such as Gregory Paul, acknowledge that the basal dinosaurs such as *Herrerasaurus* showed no bone pneumaticity that would argue for the air sac system and that, if it were present, air sacs in these first dinosaurs would necessarily have been abdominal, thus leaving no fossil record. The evidence for or against air sacs in dinosaurs comes down to rib morphology, rather than bone pneumaticity. As we have seen, the fantastic air sac system in birds can work because their ribs are highly mobile, being jointed or hinged in such a way that would allow the required pattern of avian respiration. Another possible breathing pattern has been proposed, one in which the dinosaurs used dermal ossifications in the abdominal wall to ventilate an avian-style air sac system. This kind of breathing pattern, called cuirassal breathing, could be used to inflate any air sacs and thus might be evidence that air sacs were present. But again, this system, if it were present, seems less developed in the first bipedal saurischians than in later ones.

Where does this lead us? What systems would be optimal and what systems seem allowable by the osteological evidence of the first dinosaurs? First, metabolism. A warm-blooded dinosaur would have an advantage over a cold-blooded dinosaur—in our world. All modern (cold-blooded) reptiles have to warm up at the start of the day, and thus there is little early-morning activity, other than behavioral movement, in order to acquire heat from the external environment. If the first bipedal dinosaurs—all predators—did not have to do this, they would have been able to forage freely on the slower ectotherms in the cooler morning or nighttime hours. But what is the price for this? At rest, all endo-therms use as much as 15 times the amount of oxygen as do ectotherms (there is a 5 to 15 times range based on experimental observation). In our oxygen-rich world this is not a problem for the warm-blooded animals. So much oxygen is available that there is no penalty. But in the oxygen-poor mid-Triassic, such was surely not the case. And the energy and oxygen necessary for endothermy would not have been necessary if the dinosaurs moved toward large size. With larger body size, the ratio of surface area (from which heat is lost) to body volume becomes increasingly favorable. Truly large-sized animals could have remained essentially homoeothermic in their environment even during cooler nighttime temperatures. While a lizard rapidly loses body heat in a cooler night, a 100-pound reptile does not. And the conditions of the Triassic may have been such that, thanks to highly elevated carbon dioxide levels, greenhouse heating may have kept the temperatures virtually equal day and night—and hot to boot. The Triassic climate was one suited for reptiles—hot. That heat would actually have been a problem for very large endotherms. Large dinosaurs (greater than a ton, such as most sauropods) would have overheated in even moderate temperatures, and the Triassic environment was anything but moderate. So here it is suggested that cold-blooded dinosaurs would have been a condition actually more favorable than warm-bloodedness for dinosaurs, mainly because of the large difference in oxygen needed while at rest. In the Triassic and into the Jurassic, ectothermy would not have relegated dinosaurs to a sluggish life style. With the absence of nasal turbinals characteristic of modern-day endotherms, the case for ectothermy is stronger than that for endothermy.

And what of lungs? The optimal system would have been the air sac system. Its obvious superiority in today's low-oxygen environments (at high altitude) would let any animal with such a system have greater oxygen delivery per breath compared to any other kind of lung. But there had to have been a long evolutionary history to arrive at this condition, and since crocodiles, which shared a common ancestor with the first dinosaurs, show no adaptations along this line, the evidence at hand does not support air sacs in the first dinosaurs—rather just the opposite. Neither does the osteological evidence support the hepatic piston respiration favored by John Ruben. The most parsimonious conclusion is that the first dinosaurs and perhaps all dinosaurs, had septate lungs—perhaps highly evolved and efficient septate lungs with subdivisions beyond anything extant today, but septate lungs rather than air sac lungs nevertheless. Further adaptation to low oxygen would have been accomplished by a four-chambered heart with complete separation of venous and arterial blood and perhaps by more red blood cells and other physiological adaptations. Perhaps, too, some of the peculiar bone structure observed by dinosaur bone experts is evidence of larger marrow regions involved in red blood cell formation.

What are we left with? A kind of animal unknown on Earth today. An ectotherm with phenomenally rapid growth rates and a lung system that, while inferior to the best of modern-day birds, was more efficient at extracting the thin oxygen available than those of other denizens of the day. Superiority of the dinosaurs in the latest Triassic and then into Jurassic through the Cretaceous was made possible by being better than everyone else. A return to those times might be surprising indeed, with animals showing behavior that is not mammalian or avian, not a sluggish existence but something in between. Perhaps the earliest dinosaurs were something like lions, sleeping 20 hours a day to conserve energy because of the low oxygen, but when hunting doing so actively, more actively than any of their competitors, which would have included the nondinosaur archosaurs, the cynodonts, and the first true mammals. All they needed to be was better than the rest. Clearly they were.

The question of metabolic type may also be compromised by the terms used to describe various possibilities. Dinosaur icon Jack Horner certainly thinks so. Metabolic complexes may have been far more diverse than our simple subdivision into "endothermy" and "ectothermy." While modern birds, reptiles, and mammals are put into one of these two categories, Greg Paul notes that there are many kinds of organisms that can generate heat in their bodies without external heat sources. He includes large flying insects, some fish, large snakes, and large lizards in this camp. Such animals are endotherms but not in the mammalian or avian sense. There may have been many kinds of metabolism in the great variety of dinosaurs that existed.

## BACK TO THE SEA

There are other clues than dinosaur bones to the nature of life on Earth and the challenges it faced during the low-oxygen times of the Triassic. Part of the Triassic Explosion was a diversification of reptiles returning to the sea. Many separate lineages did this, and the reasons why this happened may be tied up in the problems posed by the hot low-oxygen Triassic world.

Until now we have stressed respiratory adaptations in various animals. As we have seen, the kind of respiration used by an animal has consequences far beyond simply acquiring oxygen. Oxygen is necessary to run metabolic reactions in animals; it enables the chemical reactions that are life itself. But as in a chemistry experiment, several factors control the reactions themselves. One of the most important is temperature. Metabolic rate is the pace at which energy is used by an organism. It is far higher in endotherms than in ectoderms. But even in the same organism, the metabolic rate is directly and importantly influenced by temperature to a surprising degree. In a 2005 review of animal metabolism, physiologist Albert Bennett of the University of California at Irvine noted that one-third to one-half of all energy expenditure by an animal is used simply for staying alive through activities such as protein turnover, ion pumping, blood circulation—and breathing. Other required activities, such as movement, reproduction, feeding, and so forth are in addition to this baseline energy expenditure. If metabolic rates go up, so too does the need for oxygen, for the chemical reactions of life are oxygen-dependant. The key finding is that metabolic rates double to triple with each 10-degree temperature rise. The consequences of this in a world that had less oxygen available than now but warmer average temperatures would have been major. Thus, it has been argued here that the large dinosaurs were ectothermic, thus enjoying the best of all worlds (or at least making the best of a very, very bad world, one with no modern counterpart). The enemies were low oxygen and high heat.

There is no direct link between oxygen levels in the atmosphere and temperature. But there is a direct link between temperature and carbon dioxide, the well-known greenhouse effect. And, as we saw in Chapter 2, levels of oxygen and atmospheric carbon dioxide are roughly inverse: when oxygen is high, carbon dioxide is low and vice versa (but not without exception). Thus, there were many times in the past with low oxygen and high carbon dioxide, and thus it was hot. What a double whammy! In a low-oxygen world that is hot, animals lose in both ways—they need more oxygen than in a cool world to run their now faster metabolic reactions but have less oxygen available in the atmosphere!

We have seen many solutions to deal with low oxygen. One is obviously the simple solution of staying cool. Some solutions to staying cool, or cool enough, are physiological; some are behavioral. One of these is morphological, physiological, and behavioral: it is to return to the sea, for even in the hottest world of the past, the ocean would be essentially cooler in terms of physiology. For this reason, perhaps, many Mesozoic land animals traded feet for flippers or fins and returned to the sea at a prodigious rate.

*Hypothesis 8.2: In times of higher global temperature but lower atmospheric oxygen, an increasing proportion of tetrapod diversity is composed of animals that re-evolved a marine life style.*

Who has not been struck by the wonders of the Mesozoic, at the marvels posed by those tetrapods that returned to the sea? In the Tri-

*Reconstruction of the large Triassic ichthyosaur* Cymbospondylus. *The ichthyosaurs represented the most extreme body plan change for a once terrestrial group that returned to the sea. The low oxygen on land may have been a major impetus of this.*

assic there were giant ichthyosaurs and seagoing tetrapods such as placodonts; in the Jurassic the ichthyosaurs remained and were joined by a host of long- or short-necked plesiosaurs; and in the Cretaceous the ichthyosaurs disappeared, to be replaced by large mosasaurs. The existence of marine tetrapods is no surprise in our whale-enriched world, but the surprise to me was the sense that there were so many kinds back then. This suspicion was finally confirmed with the important research of marine reptile expert Nathalie Bardet, who in 1994 published a review of all known marine reptile families of the Mesozoic.

My reaction back then was a simple "Yes, there were lots of them," with no further interest at that time. But we returned to this data set when my colleague and research partner Ray Huey suggested to me that the high heat of the early Triassic through Jurassic would have been an evolutionary incentive for some number of reptiles to go back into the sea. We can now test this hypothesis using new data on the number of dinosaurs and the number of marine reptiles (as seen in the previously unpublished graph below). This graph indicated that there is a very interesting and inverse correlation between Mesozoic oxygen levels and the number of marine reptiles. When oxygen was low, the percentage of marine reptiles was high. But as oxygen rose, the proportion of tetrapod families that were fully aquatic markedly dropped. It may not be that the absolute number of marine forms decreased as

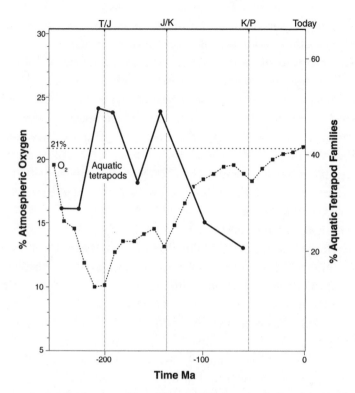

*Percent of atmospheric oxygen (dotted lines and black squares) plotted against the percentage of marine Mesozoic tetrapod families (black lines and black circles: those vertebrates that returned to the sea, such as ichthyosaurs, plesiosaurs, and mosasaurs). This graph supports the hypothesis suggested here that the number of vertebrates returning to the sea increased during high-temperature, low-oxygen times. The cooler water would have enhanced survival in a low-oxygen world. This regression analysis of these data is highly significant ($R^2 = .52$). Previously unpublished.*

much as it was that the number of terrestrial dinosaurs markedly increased. The figure demonstrating these results is shown above.

## FROM LOW-OXYGEN AIR

Let's now return to the question of why there were dinosaurs. This question can now be answered in multiple ways. There were dinosaurs because there had been a Permian mass extinction, opening the way for new forms. There were dinosaurs because they had a body plan that was highly successful for Earth during the Triassic. But perhaps these generalizations do not cut to the heart of the matter.

Chicago paleontologist Paul Sereno, who has unearthed some of the oldest dinosaurs and has made their ascendancy a major part of his study, looks at the appearance of dinosaurs in another way. In his 1999 review *The Evolution of Dinosaurs*, he noted: "The ascendancy of dinosaur on land near the close of the Triassic now appears to have been as accidental and opportunistic as their demise and replacement by therian mammals at the end of the Cretaceous." Sereno suggested that the evolutionary radiation following the evolution of the first dinosaurs was slow and took place at very low diversity.

This is quite unlike the usual pattern seen in evolution when a new and obviously successful kind of body plan first appears. Usually there is some kind of explosive appearance of many new species utilizing the new morphology of evolutionary invention in a short period of time. Not so with the dinosaurs. Sereno further noted that the dinosaurian radiation, launched by 1-meter long bipeds, was slower in tempo and more restricted in adaptive scope than that of therian mammals.

For millions of years, then, dinosaurs and other land vertebrates remained at relatively low-standing diversity, a finding that Sereno and others continue to find perplexing. In my view, this question can now be answered. Earlier we showed that there appears to be a correlation between atmospheric oxygen and animal diversity: times of low oxygen saw, on average, lower diversity than times with higher oxygen. It appears that the same relationship held for dinosaurs. To formalize this:

**Hypothesis 8.3: Dinosaur diversity was strongly dependent on atmospheric oxygen levels, and the long period of low dinosaur diversity after their first appearance in the Triassic was due to the extremely low atmospheric oxygen content of the late Triassic.**

Support for this hypothesis comes from our previous analysis of animal diversity in times of low oxygen versus high oxygen. Low-oxygen times apparently stymied the formation of many individuals (while at the same time stimulating experimentation with new body plans to deal with the bad times). This relationship has been demonstrated for marine animals, and it seems to hold for dinosaurs and other vertebrates as well. Let's examine further evidence in support of this hypothesis in the next section.

## ON THE NUMBER OF DINOSAURS

Support for this hypothesis can be found by comparing two new data sources: the latest GEOCARBSULF results for the Triassic through Cretaceous and a new compilation of dinosaur diversity through the same sampling period.

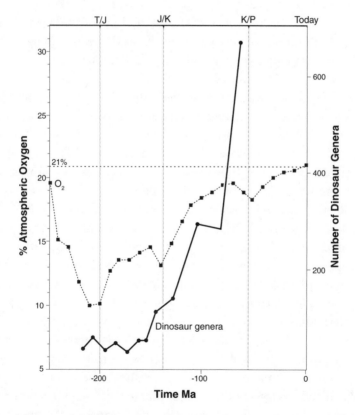

*Atmospheric oxygen percentage plotted against number of dinosaur genera. This figure supports the hypothesis that higher oxygen supported a higher diversity of dinosaurs. Part of the reason for this may be due to the fact that rising oxygen levels opened up more habitable areas at altitude, a prediction from Huey and Ward, 2005. Previously unpublished.*

Recent work by David Favkosky and colleagues, published in 2005, has provided our best estimate yet of the number of dinosaurs during the three periods of the Mesozoic. They show dinosaur genera staying roughly constant from the time of the first dinosaurs in the late middle and late Triassic through most of the Jurassic. While surely mitigated by collecting biases, the sheer number of identified dinosaur skeletons probably ensures that their overall trend is real.

It is not until the latter part of the late Jurassic that dinosaur numbers started to rise significantly, and this trend continued inexorably to the end of the Cretaceous, with the only (and slight) pause in this rise coming in the early part of the late Cretaceous. This slight drop may be due to the very small number of known localities of this age yielding dinosaurs. By the end of the Cretaceous (in the Campanian Stage) there were many more dinosaurs than during the Triassic to upper Jurassic. What was the cause of this great increase?

The figure above certainly suggests that changing oxygen levels were coincidental with changing dinosaur diversity. This is probably more than coincidence. Through the late Triassic and first half of Jurassic, dinosaur numbers were both stable and low. While originating in the latter part of the Triassic, they stayed relatively few in number until a moderate rise at the end of the period—a rise that seemed to coincide with the end-Triassic mass extinction itself. Gradually, if the oxygen results from GEOCARBSULF are even approximately correct, oxygen rose in the Jurassic, hitting 15 percent or more in the latter part of the period. It was then that the number of dinosaurs really began to increase. It was also at this time that the sizes of dinosaurs increased, culminating in the largest dinosaurs that ever evolved appearing from the latest Jurassic through the Cretaceous. Oxygen levels steadily climbed through the Cretaceous and so too did dinosaur numbers, with a great rise found in the late Cretaceous, the true dinosaur heyday. There were surely many other reasons for this Cretaceous rise. For instance, in mid-Cretaceous times the appearance of angiosperms caused a floral revolution, and by the end of the Cretaceous the flowering plants had largely displaced the conifers that had been the Jurassic dominants. The rise of angiosperms created more plants and sparked insect diversification. More resources were available in all ecosystems, and this may have been a trigger for diversity as well.

### TRIASSIC-JURASSIC MASS EXTINCTION

Let's finish this chapter with one last bit of evidence pertaining to the kind of respiration present in late Triassic animals. One of the striking new findings of the latest GEOCARBSULF has been the level of Triassic oxygen. Only several years ago the minimum oxygen levels of the past 300 million years were universally pegged at the Permian-Triassic boundary of 250 million years ago. But that time of low oxygen has been substantially moved and now may correspond more closely than previously thought with the Triassic-Jurassic boundary of 200 million years ago. Thus, we are confronted with the possibility that oxygen was lower in the late Triassic than in the early part of the period—perhaps as low as 13 percent of the atmosphere at sea level, or much less than modern-day levels. This time corresponds to one of the major changes of the Triassic: the winnowing out of most land vertebrates, with the exception of the first dinosaurs, the saurischians—creatures with pneumatized bones. This realization comes from new data compiled for this book by one of my grad students, Ken Williford, who very painstakingly went through the literature describing the various Triassic vertebrates and their ranges. The results of this long search are shown in the diagram below.

In the figure below, the Triassic-Jurassic mass extinction, one of the five most deadly mass extinctions of the past 500 million years, is shown as the horizontal black line in the middle of the figure. Here the saurischian (dinosaurs) are considered to have had air sacs.

The data gathered for this figure showed that every group *except the saurischian dinosaurs* was undergoing reduction (or at best, maintaining roughly equal diversity) in the time intervals leading up to and after the Triassic-Jurassic mass extinction. The groups with the simplest lungs (amphibians and thecodont reptiles) fared the worst and many groups that had been very successful early in the Triassic, such as the phytosaurs, underwent complete extinction. Both amphibians and thecodonts probably had very simple lungs inflated by rib musculature only. Mammals and advanced therapsids of this time, probably both having diaphragm-inflated lungs, did better, but crocodiles, presumably with abdominal pumps, did poorly. The success of the saurischians may have been due to a multitude of factors (food acquisition, tem-

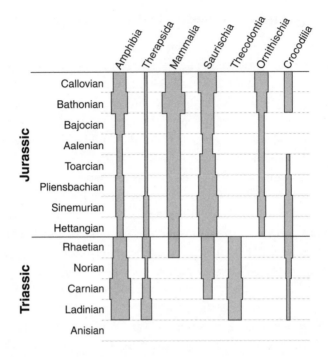

*Diversity of tetrapod genera from the middle part of the Triassic to the middle part of the Jurassic, a time interval of some 60 million years, with hypothesized breathing systems for tetrapod orders. Note that of all the orders shown, only the saurischian dinosaurs increase in diversity leading up to and after the Triassic-Jurassic mass extinction. Their then unique air sac system may have been the prime reason for this. Previously unpublished.*

perature tolerance, avoidance of predators, reproductive success), but my conclusion is that this group was unique in possessing a highly septate lung *with air sacs* that was more efficient than the lungs of any other lineage and that in the very low-oxygen world that occurred both before and after the Triassic-Jurassic mass extinction, this respiratory system conveyed great competitive advantages. Under this scenario, the saurischian dinosaurs took over the planet at the end of the Triassic and kept that dominance well into the Jurassic because of superior activity levels, which was related to superior oxygen acquisitions.

We now know that, alone among the many kinds of reptilian body plans of the middle to late Triassic, the saurischian dinosaurs diversified in the face of either static, or more commonly reducing, numbers in the other groups. We also know that oxygen reached its lowest levels of the past 500 million years in the late Triassic. Something about saurischians enhanced their survival in a low-oxygen world. In another conversation with John Ruben, I was told that this only shows that the abdominal pump allowed this enhanced survival. But what about the bone pneumaticity shown only by the group that prospered through a very bad time in Earth's history (at least if you were an air breather). And why did the other groups supposed to have had abdominal pumps do so badly—such as the phytosaurs, which went completely extinct?

The ground truth suggests that a long and slow drop in oxygen culminated in the Triassic mass extinction but that this extinction was really a double event, separated by between 3 and 7 million years. There are few places on land where this time interval with abundant vertebrate fossils can be found. We really do not know the pattern of vertebrate extinction as well as we do for the extinction in the sea. We do not know how rapidly the prominent vertebrate victims of the mass extinction such as the phytosaurs, aetosaurs, primitive thecodonts, tritylodont therapsids, and other large animals disappeared. But by the time that the gaudy Jurassic ammonites appeared in the seas in abundance, leaving behind an exuberant record of renewal in early Jurassic rocks, the dinosaurs had won the world.

What kind of lungs did they have? Here it can be proposed that they had lungs and a respiratory system that could deal with the greatest oxygen crisis the world was to know in the time of animals on Earth. Let's formalize this:

*Hypothesis 8.4: Saurischian dinosaurs had a lower extinction rate than any other terrestrial vertebrate group because of a competitively superior respiration system—the first air sac system.*

Support for this hypothesis comes from the new anatomical findings of an air sac system in saurischians and the newly presented data (above) that saurischians were actually expanding in number across this mass extinction boundary. They emerged—perhaps gasping, but still standing, over the corpses of the simpler lunged—in a world we will look at in the next chapter.

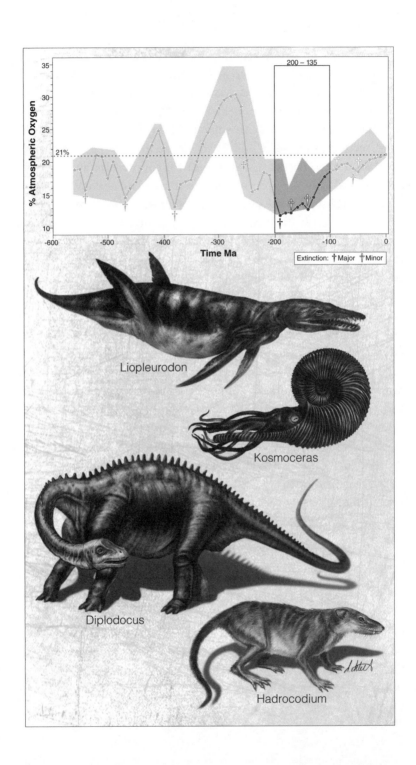

**% Atmospheric Oxygen** vs **Time Ma**

21%

200 − 135

Extinction: † Major ⊹ Minor

Liopleurodon

Kosmoceras

Diplodocus

Hadrocodium

# 9

## THE JURASSIC:
## DINOSAUR HEGEMONY IN
## A LOW-OXYGEN WORLD

S ome 200 million years ago, only 50 million years after the great paroxysm of the Permian extinction, the Triassic Period came to an end in another blood letting. As we saw in the previous chapter, of the many lineages of land life that suffered through this extinction, it was only the saurischian dinosaurs that came through unscathed. The mass extinction ending the Triassic Period was not just a phenomenon on land. It also wiped out most stocks of chambered cephalopods, but in the lower Jurassic they rediversified in three great lineages: nautiloids, ammonites, and coleoids. Scleractinian reefs flourished once again, and large numbers of flat clams colonized the seafloor. Marine reptiles belonging to the ichthyosaur and the new plesiosaur stocks again were top carnivores. On land the dinosaurs flourished, and mammals retreated in size and numbers to become a minor aspect of the land fauna but showed a significant radiation into the many modern orders near the end of the Cretaceous. Birds evolved from dinosaurs in the latter parts of the Jurassic. This is all well known and not the topics of the sort of revisionist history that is the goal of this book. Instead, let's look at the record of oxygen during the Jurassic and compare that to the numbers and kinds of dinosaurs in the ancient Jurassic Park.

## THE JURASSIC BESTIARY

As we saw in Chapter 8 the first dinosaurs were the bipedal saurischians, and they soon spawned another group, the ornithischians. The ornithischians, which began as relatively small, carnivorous bipeds, quickly evolved into herbivores and stayed that way, with both quadrupeds and bipeds. But they remained a relatively small part of the terrestrial fauna during the first half of the Jurassic. The reason for their rarity relative to the number of saurischians is probably related to oxygen levels. Unlike the saurischians, which showed bone pneumatization, *ornithischians never evolved this trait.* If bone pneumatization is a consequence of the air sac respiratory system, it can thus be inferred that ornithischians never used this kind of lung. Dinosaur expert Robert Bakker disagrees, suggesting that all dinosaurs used the air sac system, but that it was less developed in ornithischians, having only the abdominal sacs and not the sacs that fit into cervical bones in the neck. It seems more likely that the ornithischians had no air sacs at all.

The Jurassic world was not solely a dinosaur habitat, of course. We know that the Jurassic and the succeeding Cretaceous Period were times of innovations on land, in the sea, and also in the air, and it was the expansion into the air by two major groups of vertebrates that might have been the most radical of all Jurassic evolutionary changes. Three distinct kinds of flyers were in the skies: pterodactyls, pterosaurs, and birds, with the last evolving from saurischian bipeds during the latter parts of the Jurassic. While no one disputes that true birds first appeared in the Jurassic, there is still controversy about when these earliest birds took to the air. Most experts believe that the first birds could fly. Others, such as John Ruben, consider that true flying did not occur until the early Cretaceous. Regardless, true birds were on the scene in the Jurassic, and once again it appears that this extreme evolutionary novelty was a consequence, or was stimulated by, lower-than-now oxygen conditions that existed until near the end of the Jurassic itself.

Finally, the Jurassic (continuing into the Cretaceous) was the time when several lineages of true mammals evolved, all at small size, while in the seas the ichthyosaurs competed with both long- and short-necked plesiosaurs.

Thus, the dinosaurs were not alone on the Jurassic stage. But they were certainly dominant on land. Let's break their history down into distinct stages and then attempt to explain why these stages came about.

## SIX STAGES IN THE HISTORY OF DINOSAUR FAUNAL MAKEUP

While at first it seems that there were many kinds of dinosaur body shapes, in fact there were really but three: bipeds, short-necked quadrupeds, and long-necked quadrupeds. All three shared a common characteristic with birds and mammals—a fully upright (rather than sprawling) posture. Let's look at these shapes in the context of atmospheric oxygen and ambient temperature when they originated and when they thrived.

Any history of life must come to grips with and explain the extreme gigantism of the Jurassic dinosaurs. Iconic of these dinosaurs are the sauropods, the long-necked, long-tailed saurischians that are synonymous with dinosaur in most of popular culture. This body plan resulted in the largest sizes of land life ever evolved, and this begs the question, why? Why this shape and why so big? Why did other groups of dinosaurs not converge on this shape? Let's look for possible explanations for that dinosaur body plan and gigantism in relation to the prevailing oxygen content.

The history of dinosaurs can be summarized as follows:

*Middle Triassic* The earliest dinosaurs appeared in the last third of the Triassic but remained at low diversity for their first 15 million years. The majority of forms were bipedal, carnivorous saurischians. Toward the end of the period, quadrupedal saurischians (sauropods) evolved. Ornithischians diverged from the saurischians before the end of the Triassic but made up a very small percentage of dinosaur species and individuals. For much of the Triassic, dinosaur size was small, from 3 to 10 feet in length, while the earliest ornithischians (such as *Pisanosaurus*) were 3-feet-long bipeds that had a new jaw system specialized for slicing plants.

*Late Triassic* In the latest Triassic the first substantial radiation of dinosaurs occurs. It took place among saurischians, with the evolution of both more and larger bipedal carnivores and the first gigantism among early sauropods (such as *Plateosaurus* of the Upper Triassic).

*Early to Middle Jurassic* Continuation of the trend of the latest Triassic, where saurischian bipeds and quadrupeds dominated faunas, characterized this phase. During this time, however, the ornithischians, while remaining small in size and few in number, diversified into the major stocks that would ultimately dominate dinosaur diversity later in time, in the Cretaceous. These stocks included the appearance of heavily armored forms (such as the thyreophorans). These were quadrupeds and included the first stegosaurs of the middle Jurassic. A second group was the unarmored neornischians (which included ornithopods, hypsilophondontids, iguanodons, and duck bills), marginocephalians (the ceratopsians, which did not appear until the Cretaceous), and bone-headed pachycephalans. But it was the sauropods that were most evident in numbers. They split into two groups in the latest Triassic, the prosauropods and true sauropods, and in the early and middle Jurassic the prosauropods were far more diverse than sauropods but went extinct in middle Jurassic time, leading to a vast radiation of sauropods into the late Jurassic.

Finally, the bipedal saurischians also showed diversity and success in the lower and middle Jurassic. In the latest Triassic time they split into two groups (the ceratosaurs and tetanurans). The ceratosaurs dominated the early Jurassic, but by middle Jurassic time the tetanurans increased in number at the expense of the ceratosaurs. They too split into two groups, the ceratosauroids and the coelophysids. The latter group eventually produced the most famous dinosaur of all, the late Cretaceous *Tyrannosaurus rex*, although its middle Jurassic members were considerably smaller. Their most important development in the Jurassic was evolution of the stock that gave rise to birds.

*Upper Jurassic* This was the time of the giants. The largest sauropods came from late Jurassic rocks, and their dominance continued into the early part of the Cretaceous. Keeping pace with this large size were the saurischian carnivores, with giants such as *Allosaurus* typical. Thus, the most notable aspect of this interval was the appearance of sizes far larger than in the early and middle Jurassic. This was not only among the saurischians. During the late Jurassic the armored ornithischians also increased in size, most notably among the heavily armored stegosaurs. The diversification of ornithischians at this time—with the appearance of stegosaurs, ankylosaurs, nodosaurs, camtosaurs, and hypsilophontids—radically changed the complexion of the dinosaur assemblages.

*Early to Middle Cretaceous* While the dominants for the early part of this interval remained large sauropods, as the Cretaceous progressed, a major shift occurred: ornithischians increased in diversity and abundance until they outnumbered saurischians. Sauropods became increasingly rare as many sauropod genera went extinct at the end of the Jurassic.

*Upper Cretaceous* Dinosaur diversity skyrocketed. Most of this diversification came through large numbers of new ornithischians: ceratopsians, hadrosaurs, and ankylosaurs, among others. Only a small number of sauropods were present.

So how to make sense of this pattern? In his review of the dinosaurs, Paul Sereno noted the difference between any of the separate dinosaur diversifications and mammals after the Cretaceous. One great difference, of course, is that the Cretaceous-Tertiary extinction was unique in being so catastrophic and so short in duration. The end-Triassic mass extinction, for example, was a far more extended event. Nevertheless, Sereno's observations are important, and in his view, crying out for explanation. For example, he noted:

> The radiation of the nonavian dinosaurs, by comparison to the Paleocene mammals was sluggish and constrained. Taxonomic diversification took place at a snail's pace; standing diversity, which may have totaled 50 genera or less during the first 50 million years (the late Triassic and early Jurassic) increased slowly during the Jurassic and Cretaceous, never reaching mammalian levels; maximum body size for herbivores and carnivores was achieved more than 50 million years after the dinosaurian radiation began, only 8 to 10 distinctive adaptive designs evermore and few of these would have been apparent after the first 15 million years of the dinosaur radiation.

Sereno thus found no reasonable explanation for this early dinosaur history, and most other workers have yet to even take the emergence into account in any meaningful, hypothesis-driven way. Here, however, we can propose several hypotheses that readily explain the history above, and the pattern of subsequent dinosaur history. For any reader who has made it this far, the explanation will be no surprise: the history of dinosaurs was largely dictated by changing oxygen levels.

## HYPOTHESES FOR THE OBSERVED HISTORY OF DINOSAUR DIVERSITY, DISPARITY, AND SIZES

Here we can propose the following hypothesis based on dinosaur morphological and taxon abundance in the context of a long-term rise in Mesozoic oxygen levels. These explanations can answer the questions posed by Sereno about the observed history of dinosaurs. They are as follows.

*Hypothesis 9.1: The Carnian through Hettangian interval (late Triassic to earliest Jurassic) was a time of low-oxygen levels and this coupled with very high carbon dioxide levels and hydrogen sulfide poisoning—not asteroid impact—was the major cause of the Triassic-Jurassic mass extinction.*

Support for low-oxygen levels over this interval comes from marine stratal evidence of progressive anoxia (black shales, laminated beds, and trace fossils [such as chondrites] characteristic of low-oxygen sea bottoms), from the modeling of GEOCARBSULF, and from studies of fossil plants across this time interval by University of Chicago paleobotanist Jenny McElwaine, who showed morphological evidence of rising carbon dioxide in her fossil plants collected over this time interval. The combination of low oxygen, high global temperatures, and, based on the recovery of new biomarkers, perhaps hydrogen sulfide was the killing mechanism. While there were successive asteroid impacts over this time interval, one of which (the Manicouagan event of 214 million years ago, or 14 million years before the Triassic mass extinction) was large, the impacts played little or no role in the extinctions according to scientific observations of many stratal sections of this time interval.

*Hypothesis 9.2: Ornithischian dinosaurs did not possess as effective a respiratory system as did saurischians. However, they were competitively superior to herbivorous saurischians with regard to food acquisition. With the rise of oxygen to near present-day levels in the Cretaceous, ornithischians became the principal herbivores because of this superiority, leading to the extinction of many saurischian herbivores through competitive exclusion.*

While Robert Bakker has argued that ornithischians and saurischians alike had an air sac respiratory system, the lack of bone pneumaticity in any of the ornithischians indicates that if they used an air sac system it would have been rudimentary. What they did possess was a series of tooth adaptations that may have been far superior to the teeth in herbivorous saurischians. While we have seen that the Jurassic to Cretaceous interval marked a relatively rapid and significant rise in atmospheric oxygen, other events were taking place, including a radical change in flora. Dinosaurs evolved in a gymnosperm-dominated world—with conifers, but with ferns, cycads, and gingkoes as well. But in the early part of the Cretaceous a new kind of plant appeared, a flowering plant. With this new kind of reproduction and other adaptations, these plants, the angiosperms, underwent a rapid adaptive radiation. They out-competed the earlier flora nearly everywhere on Earth to the extent that by the end of the Cretaceous, some 65 million years ago, the angiosperms made up as much as 90 percent of vegetation. This transition in available food types would have affected the herbivores, and the kind of herbivores available as food would have directly affected carnivore body plans. Killing a late Jurassic sauropod would have been very different from killing a late Cretaceous hadrosaur.

Herbivory is dependent on the correct kind of teeth for the available plants. The sauropods may have lived on pine needles, their huge barrel bodies being, essentially, giant fermenting tanks for digestion of a relatively indigestible food source. The appearance of broadleafed plants, the angiosperms, would have required different teeth and biting surfaces than those optimal for slicing pine needles off trees. Thus, the transition from the sauropod-dominated faunas of the Jurassic to the ornithischian-dominated faunas of the Cretaceous was surely related in some part to the change in plant life. But respiration may have played a part as well, and perhaps if oxygen had not risen above 15 percent, the ornithischian takeover would not have taken place.

## JURASSIC-TRIASSIC DINOSAUR LUNGS AND
## THE EVOLUTION OF BIRDS

In Chapter 8 we looked at the questions of dinosaur metabolism and respiration. The proposal there was that the first dinosaurs were of a kind of animal never seen and not alive today. Through upright posture and an evolving air sac system, they developed respiratory efficiency (the amount of oxygen extracted from air per unit time, or per unit energy expended in breathing) superior to any other then-extant animal. But these early forms may have lost (or never gained in the first place) endothermy, replacing it with a more passive homothermy or even ectothermy, which was attained with larger size. That was their trick—using ectothermy to reduce oxygen consumption while at rest and a superior lung system to allow extended movement without going into rapid anaerobic (and thus poisonous) states. But what of the later dinosaurs? We know that birds, a group of dinosaurs first appearing in the Jurassic, eventually had both endothermy and a very different kind of lung than in any extant reptile. It seems probable that the large and small saurischians parted company, with smaller forms evolving endothermy later in the Jurassic as oxygen levels rose rapidly.

John Ruben's group stakes out a very different and conservative position that the first true birds had both ectothermy and reptilian, not air sac, lungs. Most other dinosaur and bird specialists are not so sure. Some believe that endothermy and air sac lungs of some kind were present in *Archaeopteryx*, while others indicate that based on bone pneumaticity the air sac system was present in the bipedal carnivores that gave rise to birds.

With perhaps the exception of the always-fascinating tyrannosaurids, no group of dinosaurs has received more attention in recent times than the basal birds. Vigorous debate centers on their body covering and, most importantly, on when flight first evolved and why. The first birds appeared about 150 million years ago, and the famous first bird remains *Archaeopteryx*. That is just before the start of the Cretaceous. Oxygen had been rising for 50 million years at that time. Gigantism in dinosaurs was common. The immediate ancestors of the birds were fast, ground-running dinosaurs that may have used their forelimbs for a type of predation, a motion that was preadapted for a wing stroke in a flyer, according to Berkeley paleontologist Kevin Padian. The fossil record suggests that the ancestors of the first bird were the bipedal carnivorous saurischians known as troodontids or perhaps the dromaeosaurids, forms that appear to have been already feathered (there is much controversy about this).

Could *Archaeopteryx* fly? Padian thinks so. But there is debate about when true flight took place. Could the late Jurassic "birds" really fly, at a time when their competition in the air would have been the diverse and successful pterodactyls? The fossil record does show that by the lower Cretaceous there was a bird (*Eoaluolavis*) that had evolved a "thumb wing," an adaptation that allows greater maneuverability at slower speeds. Thus, within a few million years after *Archaeopteryx*, fairly advanced flight was present. New discoveries from China have revealed an unexpected high diversity of birds by the early part of the Cretaceous. Flight was an adaptation that stimulated a rapid evolution of new forms.

What new information can be added? Flight is highly energetic. Birds use a great deal of energy to fly, and that, added to their relatively small size and endothermy, makes them great users of oxygen. So the air sac system serves them well. But what of the late Jurassic, when oxygen may have been somewhat lower than now? Could it be that the even lower oxygen of the early and middle Jurassic delayed flight? What about other known flyers? The pterodactyls had long been in the air by the evolution of the first bird, but pterodactyls may not have been as energetic in their flying or might have had an air sac system, for they also show bone pneumaticity consistent with the presence of an air sac system. Hence, there are questions about lung type not only in the first birds but also in other flyers, their immediate ancestors, and in the bipedal, saurischian dinosaurs like *T. rex* that came along in the Cretaceous.

## DINOSAUR REPRODUCTION AND OXYGEN LEVELS

Alas, the total extinction of dinosaurs 65 million years ago (unless birds are considered dinosaurs, now accepted by many) will forever make it impossible to answer many pressing questions about their biology. So it is natural that we try to answer these questions using their

nearest living relatives, the birds and reptiles. After lung type and metabolic type, some of the most interesting questions relate to reproductive strategy, and this too can be examined in the light of changing
oxygen levels.

Birds show little variation in their reproduction. Extant birds, our
best window to the dinosaurs, all lay eggs with a porous calcareous
shell. There are no live births in birds, in contrast to extant reptiles,
which have many lineages using live birth.

There is also great variation in egg morphology between birds and
some reptiles. While the eggshell in birds and reptiles consists of two
layers, an inner organic membrane overlain by an outer crystalline
layer, the amount of crystalline material varies greatly, from a thick,
calcium carbonate layer like that in birds to almost no crystalline material at all, so that the outer layer is a leathery and flexible membrane.
Even the mineralogy of the crystalline layer varies, from calcite in birds,
crocodiles, and lizards to aragonite (a different crystal form of calcium
carbonate) in turtles. Eggs are thus divided into two main types: hard
or crystalline and soft or parchment (some scientists further subdivide
the parchment eggs into flexible [used by some turtles and some liz-
ards] and soft [parchment, used by most snakes and lizards]). Not surprisingly, the fossilization potential of these different hardness
categories of eggs differs markedly. There are numerous fossil hard eggs
known (many from dinosaurs), a few flexible eggs, and no undisputed
soft eggs preserved.

Because of the great interest in dinosaurs there has been much
speculation about their reproductive habits (the thought of two gigantic *Seismosaurus* mating rather boggles the imagination), and there are
still many mysteries. One of the seminal discoveries about dinosaurs
was that they laid large calcareous eggs, with calcite crystals making up
the mineral layer, a finding from the first expedition to the Gobi Desert
by an American Museum of Natural History expedition in the 1920s.
Since then, thousands of Cretaceous dinosaur eggs have been found
around the world, and even the nesting patterns have been discovered,
the most notable being the nest discoveries in Montana by Jack Horner.

But are these Cretaceous finds characteristic of dinosaurs as a
whole? This question remains unresolved and controversial. While
most scientists assume that all dinosaurs laid hard-shelled eggs, this is

far from proven, and, as we shall see below, there is indirect evidence that some early dinosaurs may have utilized parchment eggs or even live births.

What is the egg record for dinosaurs? Almost all dinosaur eggs come from the Cretaceous, and the nature of their crystal form and size, as well as the number and pattern of pores in the egg, show a wide variety. There are certainly plenty of eggs found from the Cretaceous, but, while known, there are far fewer Jurassic dinosaur eggs and almost none known from the Triassic.

There are several possibilities for this. Perhaps there is some preservation bias, with pre-Cretaceous eggs as common as those of the Cretaceous, but the lesser extent of Triassic and Jurassic dinosaur beds compared to the vast expanse of Cretaceous-aged beds has caused this difference. Another possibility is that pre-Cretaceous eggs fossilized much less readily than those from the Cretaceous. This would certainly be the case if pre-Cretaceous eggs were leathery like those of extant reptiles, rather than calcified like birds. And if, like the marine ichthyosaurs, some dinosaurs utilized live birth rather than egg laying there would certainly be fewer eggs to find. As in so many other aspects of the history of life, the level of atmospheric oxygen may have played a major role in dictating mode of reproduction.

Fossil eggs from Cretaceous deposits attributed to dinosaurs (what else could have laid such large eggs?) have a calcium carbonate covering like a chicken egg (but thicker). But unlike chicken eggs, which are smooth, the dinosaur eggs were usually ornamented with either longitudinal ridges or nodular ornamentation. Presumably, ornamentation allowed the eggs to be buried after emerging from the female, with the ornament allowing airflow between the eggs and the burial material. The ability to bury eggs may have aided their fossil preservation potential and perhaps helps explain why there are so many Cretaceous eggs and so few other kinds. The heavy calcification would also help the eggs withstand the overpressure of burial in soil or sand. Also the complex behavior involved in nest making and orienting the eggs in burial mounds is now known for the late Cretaceous—but not before.

What are the advantages of calcareous eggs? They are strong, harder for predators to break into, and aid in development. As the embryo grows inside the egg, some of the calcium carbonate is dissolved from the eggshell itself to be used in bone growth. Eggshells also might shield the egg from bacterial infections. But this comes at a price. Calcium carbonate, even an eggshell-thin layer of it, will not allow the passage of air or water into or out of the shell. But developing embryos need both water and oxygen. All calcareous eggs thus have pores so that oxygen-laden air can enter but not so many pores that water quickly leaves by desiccation. To ensure sufficient water, the interior of the egg has a large amount of a compound known as albumin (familiar to us as the "white" of a chicken egg), which provides water to the embryo. This kind of egg is found in all birds and crocodiles.

The second kind of reptilian egg, the parchment egg, is found in turtles and most lizards. This kind of egg can take up water and actually expand in size with water uptake. But water permeability is a two-way street: parchment eggs can easily lose water too. Burying these kinds of eggs in nests, the habit of many turtles and alligators and crocodiles, reduces water loss and helps hide these eggs from predators.

Burying an egg presents some danger. All developing embryos require oxygen, and thus the embryo requires an egg that can allow the passage of oxygen from the atmosphere into the egg. If the egg is buried too deeply or in impermeable sediment, the embryo will suffocate. And if the egg is laid at high altitude, it runs the risk of the same fate, despite being smothered by parental care. So far biologists have concentrated on temperature as the major variable affecting development rates in reptiles and birds. But the clues given by high-altitude lizards suggest that oxygen levels certainly play a part as well. Lizards living at altitude often show live birth. They also hold the eggs in the birth canal for long periods of time. In both cases, the explanation has been that this is done to maintain relatively high temperatures in an environment where there can be very cold temperatures that could slow development. But both of these adaptations would lessen or completely remove the time that the embryo is enclosed in a capsule that itself reduces the rate of oxygen acquisition. Calcareous eggs cannot be held in the mother because they do not allow oxygen to enter the eggs until they emerge from the mother.

So we have a mystery. Reptiles show four different kinds of reproduction: live births, parchment eggs that are held within the mother for extended periods of time, parchment eggs that are laid soon after formation within the mother, and calcareous eggs. Following birth there is also a series of possibilities: the eggs are buried or not, and when not buried, the eggs can be cared for by the parent or not. The advantages of each of these and the time they first appeared are unknown.

And we have a second mystery. Most known dinosaur eggs are from the Cretaceous (mainly late Cretaceous) and are calcified. Also, the appearance of burial behavior in dinosaurs is also characteristic of the late Cretaceous. But what of the pre-Cretaceous dinosaurs? While there are eggs from sauropods and bipedal saurischians from the late Jurassic, most spectacularly from deposits in Portugal where the eggs contain the bones of embryos, earlier rocks are nearly barren of dinosaur eggs and/or nests. Only a single confirmed egg is known from the Triassic.

When these various kinds of eggs first evolved thus remains a mystery. In 2005, Nicholas Geist and John Ruben published an abstract in which they proposed that calcareous eggs first appeared at the end of the Permian as an adaptation to avoid desiccation in the increasingly dry late Permian through Triassic global climates. Unfortunately, there is no fossil evidence to support this: there are no accepted Permian eggs despite the existence at that time of anapsids, diapsids, and synapsids, and only a small number of late Triassic eggs that may have been from dinosaurs. Eggs commonly preserved in Cretaceous sediments, in the same kind of sedimentary environments that can be found in Permian and Triassic rocks. In all likelihood, if archosaurs used hard eggs in the Permian or Triassic, they would already have been found. The absence of evidence is always a dangerous tool, though; all too well known, and deservedly so, is the hoary adage that the absence of evidence is not necessarily evidence of absence.

In their wonderful 1997 summary of dinosaur eggs, Darla Zelenitsky and the late Karl Hirsch recognized only two dinosaur egg shapes (rounded and elongated) but seven different patterns of crystal arrangement. This diversity of egg-wall morphology would be surprising if all dinosaurs had evolved from a single, egg-laying ancestor—but would be what we would predict if hard-shelled egg laying evolved numerous times by separate lineages of dinosaurs. If we add the additional (and different) eggshell morphologies found in extant reptiles and birds, there are a combined 12 separate eggshell microstructures that have evolved. Perhaps each of these is an adaptation to a different kind of stress that each egg undergoes: a turtle egg in a deep burrow, for instance, faces a very different series of challenges than does a bird egg in a nest high in a tree. But the more likely is that hard eggs separately evolved in multiple lineages—including dinosaur lineages.

With this (admittedly tenuous) evidence at hand, a different scenario can be proposed from the Geist-Rubin hypothesis that archosaurs evolved the hard eggshell in the Permian. The extreme environmental conditions that Geist and Rubin noted—high temperature and thus desiccation—and the other deleterious factor that they didn't—low-oxygen values—would have stimulated evolutionary change from the system used by animals that had first evolved in a high-oxygen and perhaps lower-temperature world. Seemingly, a better response to heat and low-oxygen (which is further magnified by the heat) would be live birth. The evolution of live birth thus may have come about in response to lowering global oxygen values in the late Permian. Unfortunately, evidence for this is again an absence of evidence. Nevertheless, it is what we have to work with: despite the enormous number of therapsid bones found in South Africa, Russia, and South America, a fossil egg or nest has never been found in these rocks. Therapsids may have already evolved live births by this time, a trait carried on by their descendents—the true mammals that were first found about the same time as the first dinosaurs appeared on the scene.

What can we learn about oxygen and eggs in living organisms? While data are surprisingly scanty, it appears that oxygen can diffuse through the wall of a modern parchment egg more readily than it can in a calcareous egg. Parchment eggs can also be kept within the mother's body for extended periods of time, where it is maintained in an oxygen-rich environment. It may be that many lineages of dinosaurs evolved the calcareous egg in the late Jurassic as a response to rising oxygen and that the formation of calcareous eggs, which are then buried, was a reproductive strategy that was not viable in the late Permian through middle Jurassic environments of lower atmospheric oxygen.

So, let me more formally pose this:

*Hypothesis 9.3: The low-oxygen and high-heat conditions of the late Permian into the Triassic stimulated the evolution of live birth and of soft eggs that would have been effective at allowing oxygen movement into eggs and carbon dioxide out. On the other hand, the higher-oxygen levels (and continued high temperatures) of the late Jurassic-Cretaceous interval stimulated the evolution of rigid dinosaur eggs and egg burial in complex nests.*

Only time will tell if new discoveries from late Triassic through middle Jurassic strata will add significant new information to the topic of dinosaur (and mammalian!) birth strategy. Like characteristic metabolisms, the contrasting patterns of live births versus egg laying are fundamentally important—and ones that have received surprisingly scant attention by evolutionary biologists. Solving this problem—by learning the time of origin and the distribution of one kind of birth strategy or the other—should be a major research topic of the near future but, sadly, may prove to be intractable because of the non-preservation of parchment eggs.

### JURASSIC-CRETACEOUS IN THE SEAS

Let's now move from land to sea. The Jurassic and Cretaceous oceans would have been dangerous places to swim in. The major Triassic marine predators just increased in number in the Jurassic and added a new and efficient kind of fish- and cephalopod-eater, plesiosaurs. In the Cretaceous yet another kind of tetrapod predator appeared as well: mosasaurs displaced plesiosaurs and ichthyosaurs as the top carnivores in the sea. Ammonites continued to flourish but evolved large numbers of uncoiled shell shapes in addition to the traditional planispiral shapes of the Jurassic and lower Cretaceous. A diverse calcareous plankton including coccoliths and foraminifera changed the nature of planktonic communities, while a new form of gastropods, the entirely carnivorous neogastropods, joined a host of other shell-breaking predators to totally transform the benthos in what has been called the Marine Mesozoic Revolution. A response to this increase in predation was the evolution of infaunal siphonate clams (with heterodont dentition) and stalkless crinoids (the comatulids). All of these events are well known and documented and thus are not the targets of revisionist history. But the reasons behind the specific designs of some of these same organisms are another matter. The rising oxygen of the Jurassic and Cretaceous following the Triassic nadir caused an increase in the number of species, but, as we have seen, it was low oxygen that stimulated new kinds of body plan. Let's look at four types of aquatic body plans that are related to oxygen levels in the seas and that also made the Mesozoic oceans very different places than our present-day oceans.

　　1.　　*The evolution of low-oxygen-tolerant bivalves.* The oxygen low of the early through middle Triassic produced a new and poorly habitable ocean. As we have seen, animals do very badly in low oxygen. Atmospheric oxygen levels affect oceanic oxygen levels and quite often even serve to magnify the effects. Many ocean bottoms of the Mesozoic were completely anoxic, and most were at least hypoxic. Rare was an ocean community of these times that had oxygen levels like those on the bottoms of modern-day oceans.

　　Just as in the Cambrian Explosion, where animals were stimulated to produce new kinds of body plans based around respiratory systems, so too did animals of the Triassic seas show a multitude of new adaptations. As we have seen, the land fauna experimented with a variety of lung types. The same kind of exploration took place in the oceans. The bivalved mollusks were one group that evolved a new kind of body plan, and even physiology, in response to the nearly endless expanse of nutrient-rich but low-oxygen bottoms.

　　The very lack of oxygen on the ocean bottoms made them, in one sense, wonderful places to live. Vast quantities of reduced carbon, in the form of dead planktonic and other organisms, fell to the seafloor and were buried there. On an oxygenated bottom this material would soon be consumed, by filter- or deposit-feeding organisms and scavengers, and used for food. But the low-oxygen conditions kept these organisms out, and not even the usual bacteria that decompose dead creatures on the sea bottom were around. As we have seen, this is one reason that oxygen levels plummeted in the Triassic. But the clams figured a way out of this. A few kinds, living on the seafloor of the ocean bottoms that had at least some oxygen, fed not on the falling organics but on methane-containing compounds coming up from some fraction of the organic-rich sediment. Methanogens are a group of bacteria that thrive in low- or no-oxygen conditions, and even several inches down into the sediment on a sea bottom with some oxygen, they would have penetrated into an oxygen-free zone—thus it was an ideal environment for methanogens.

As methanogens metabolize, they release methane as a by-product. The Mesozoic clams may have had other bacteria in their gills that could exploit the methane and other dissolved organic material, or they may simply have fed on the bacteria. A somewhat similar mechanism is found today in the deep-sea vent faunas, where giant tubeworms and clams use these chemicals as food. But the difference is that the modern vent faunas are oxygenated. The animals down there do not even need gills. The clams of the Triassic and Jurassic were not so lucky.

These kinds of clams are found in huge numbers in Triassic and Jurassic sediments. In the latest part of the Triassic, when oxygen reached its lowest levels, the number of these clams was so great that they formed rocks themselves with their shells—a kind of rock known as a coquina. Two Triassic taxa that did this were *Halobia* and, especially, *Monotis*. Both lay on the surface of the sediment and were immobile. There was no burrowing (like the majority of today's clams) or movement on top of the sediment (like modern-day cockles). They were more akin to mussels—they just sat there. And like mussel beds, they were often abundant. In recent Triassic work, many scientists have now sampled and seen these kinds of beds from all over the world in rocks of this age, and the shock of seeing so much life packed into rocks is always striking. For tens of feet of stacked strata there is nothing but an endless packing of shells. There must have been billions of these clams lining the bottoms of the sea, presenting a spectacle that has no parallel today. But the most unusual aspect of the clam beds is that there is virtually nothing else in there with them. There may be the occasional ammonite or other mollusk, but these are rare indeed. The clam beds are essentially monotypic—composed of a single species. It is very unusual to see monotypic assemblages in the modern-day marine sea bottom, especially those in the tropics. But the Triassic world was so warm that virtually all the world was tropical, and yet there is never much diversity to these beds, which are found worldwide.

The flat clams began to diminish in the shallows first in the mid- to late Triassic but held on in deeper water until the end of the Cretaceous. The abundant Cretaceous clam *Inoceramus* lived in many environments, but giants of this kind are found in deep-water deposits. Some fossils are as large as 6 feet across, and like the Triassic clams they seem to have been neither carnivores nor herbivores, but instead they were chemosynthetic, using chemical compounds coming up from the reduced carbon–rich mud on which they lay. Ironically, it appears that they were eventually driven into extinction by rising oxygen levels. With the appearance of an ever-more oxygenated sea bottom in the upper Cretaceous, as atmospheric oxygen rapidly rose, the conditions that had succored the flat clams disappeared.

2. *The evolution of low-oxygen-tolerant cephalopods.* There are many places in the world where marine strata of the latest Triassic age are overlain by Jurassic strata. At such outcrops one can walk through time, and if the strata are continuous, the dramatic events of the late Triassic and early Jurassic are present for all to see. This interval of time and rock preserves evidence of the great Triassic mass extinction, one of the so-called Big Five mass extinctions, a dubious honor of species death. As you walk through upper Triassic beds you are first in strata packed with fossils of the flat clam *Halobia*; then you move into younger rocks with the even more abundant *Monotis*, the clam described just above. Then these clams disappear in turn, over only several feet of strata, leaving a long barren interval of rock and time, the last stage of the Triassic, an interval perhaps 3 million years in length known as the Rhaetian stage. Finally, after this thickness virtually without fossils, a new group suddenly appears—the ammonites.

While there are ammonites to be found in the upper Triassic, they are never abundant. Most seem to have gone extinct when the clams did, perhaps 3 million years before the final phase of this end-Triassic mass extinction. This situation changed drastically with the onset of the Jurassic. Most famously at the beach of Lyme Regis of England, but also in southern Germany and at many other localities worldwide, the earliest Jurassic ammonites appear in huge numbers, and they diversify over only a few short meters of strata as well. This is not like the Triassic flat clams where one species is all you get. These ammonites of the first part of the Jurassic are diverse and abundant. It is a fossil collector's dream, and it tells us that the great drop in oxygen was finally over and that a slow rise in oxygen was finally underway. But the ammonites were not telling us that oxygen levels similar to today were suddenly in place. The ammonites appear because the surface of the early Jurassic seas began to have a modicum of oxygen, and the ammonites took full advantage. They did so because they were among the best animals on Earth for dealing with low oxygen.

Chapter 3 presented a new potential pathway for the evolution of the first cephalopods. The nautiloids are still with us today—but up until the end of the Cretaceous, the nautiloids from the Devonian on were far outnumbered by one of their descendants, the ammonites.

Because of the overall similarity of the chambered shells in both nautiloids and ammonoids, we presume they may have had somewhat similar modes of life. Nautiluses today live in highly oxygenated water over most of their range. But here and there they also live in hypoxic bottoms. This was a great curiosity when first discovered, because it was conventional wisdom that all cephalopods need high-oxygen conditions. Not so the *Nautilus*. My decade of studying them proved that they are very tough and resistant when taken out of the water. They can sit out 10 or 15 minutes with no ill effects. When they are in water, they quickly replenish oxygen in their blood through one of the relatively largest and highest-powered pump gills ever evolved. If ever an animal was adapted for low oxygen, this is it.

British zoologist Martin Wells, who measured oxygen consumption of various captive nautiluses in New Guinea, finally proved this. When *Nautilus* is confronted with low oxygen, it does two things. First, its metabolism slows way down. Second, it apparently uses some of the gas in its air-filled chambers for emergency respiration.

The mass appearance of ammonite fossils in lower Jurassic strata suggests that, like the nautiluses, the ammonites were superbly designed to extract maximum oxygen from minimal dissolved volumes of the oh-so precious gas. They do so because of a powerful pump gill system that was capable of moving sufficient volumes of low-oxygen water across the gill surface to yield the necessary number of oxygen molecules from seawater to live. To formalize this:

*Hypothesis 9.4: Jurassic-Cretaceous ammonite body plans evolved near the Triassic-Jurassic boundary in response to worldwide low oxygen. Their new body plan (compared to the ammonoids that came earlier) involved a much larger body chamber relative to the phragmocone, which may have allowed for much larger gills. Because of this they had to use thinner shells, and this required more complex sutures. The sutures also allowed faster growth by increasing rate of chamber liquid removal. Within the large body chamber was an animal that could retract far into this space and that had very long gills relative to its ancestors.*

What support is there for this hypothesis that basal Jurassic ammonites were low-oxygen specialists? We know that early Jurassic ammonites are found in great numbers in otherwise animal-free strata, and we know their body chamber length increased at this time. Unfortunately, ammonite soft parts are still unknown, and we do not know if they had four gills (like *Nautilus*) or two (like modern-day squid and octopus). But from the very unstreamlined shells of the majority of early Jurassic forms, it is clear that these animals were not fast swimmers. It is far more likely that they floated slowly or swam gently near the surface, using their air-filled shell like a zeppelin. Their pump gill forced huge volumes of water across their lungs in short periods of time, allowing them to live where most animals could not.

The ammonites went on to stay very common right up to the end of the Cretaceous. My work in Spain and France in the late 1980s showed that they were killed off, very suddenly, as a consequence of the Cretaceous Chicxulub asteroid. But by the end they were living in higher-oxygen waters and their shapes changed subtly, allowing a more active and vigorous life style.

3.  *The evolution of crabs.* Another new body plan that may have arisen as a consequence of and as an adaptation for low oxygen is that of the lobsters and crabs. While the overall shrimp-like body form of crustaceans is found in Paleozoic rocks, crabs are a relatively new invention. A crab is simply a shrimp-like form in which the abdomen is tucked under the body. Fusion of the head and thorax into a heavily armored and calcified plate makes the crab a difficult nut to crack for its predators. And placement of the abdomen under this armor plating is design genius. It is the abdominal regions that are most susceptible to breakage in any predatory attack, and by eliminating this chink in their armor, the crabs rapidly rose to marine prominence. Their large claws allow them to crack open mollusk shells, among other prey; they are shell-breaking predators. Prior to this, few predators were able to break into shelled organisms. Crabs and others evolved the morphological means to render many previously impregnable skeletons vulnerable.

Thus, the accepted reason for the crab's body plan, novel as it is, relates to defense (tucking of the abdomen, thickening and increasing calcification of the head-thorax region) and offense (evolving a strong pair of jaws). But here is another: crab design came about in some part as a primary adaptation for increasing respiratory efficiency.

*Hypothesis 9.5: The crab's body plan evolved for multiple reasons but one was that it increased respiratory efficiency by putting the gills in an enclosed space under the cepahlothorax (the head-thorax) and then evolving a pump to move water over the now enclosed gills.*

The crab gill design is a marvelous way to increase water passing over the gills. Crabs evolved from shrimp-like organisms, and in these ancestors a progression toward the crab gill system can be seen. In shrimp the gills are partially enclosed beneath the animal. While covered dorsally, the gills are attached to segments and are open to water underneath.

4.  *The evolution of the calcareous plankton.* The formation of calcium carbonate—limestone, coming in the two mineral species calcite and aragonite—is affected by several factors, most importantly tem-

perature, pH, and the concentration of calcium and carbon dioxide levels in water. But there is another factor that is important as well—oxygen. In our world, most calcium carbonate formation is mediated or undertaken for adaptive reasons by organisms. Calcium carbonate is the most commonly used mineral in skeletal formation, and the majority of animals, and many protozoans and plants, have skeletons of one kind or another. While there are substantial numbers of organisms using silicon for skeletons, in numbers and biomass produced, they are far exceeded by those using calcium carbonate. All of these organisms also need oxygen for life, so in water containing low levels of oxygen there is little or no calcium carbonate produced. It was the substantial rise of oxygen levels from the Jurassic through the Cretaceous that increasingly favored the formation of calcareous skeletons by plankton.

The amount of calcium carbonate produced at any time on Earth has substantial effects on the atmosphere and chemistry of the oceans. When atmospheric carbon dioxide levels are high, the rate of limestone formation increases—but only if there is sufficient oxygen to allow the organisms making the skeletons to flourish. The only exception to this is among the group of plants that produce carbon dioxide skeletons, such as the single-celled planktonic forms known as coccolithophorids.

Today the largest amount of global calcium carbonate formation comes from oceanic organisms, both animal and plant. The coccoliths are the most important plants, but also of major importance is a group of protists known as foraminifera. The latter are relatively large amoeba-like creatures with different types existing in both the bottom sediment and free floating in the plankton. The skeletons of both accumulate on the bottom, and in portions of the ocean they produce thick deposits on the ocean floor. As this calcium-rich layer is eventually subducted during plate tectonics, it is heated and combines with other minerals. A by-product of this reaction is the formation of more carbon dioxide, which enters the atmosphere through volcanic processes. The calcium ends up in other minerals, but the carbon circulates between organic and inorganic phases.

The death, sinking, and ultimate burial of these two planktonic groups have a second geological effect. So high is the volume of living flesh in foraminifera and, especially, coccoliths that they increase the rate of burial of organic carbon. This in turn affects atmospheric oxygen levels. If large volumes of organic (and thus reduced) carbon are quickly buried, it causes a rise in atmospheric oxygen. This may have been an important cause of the rise of oxygen in the Mesozoic.

Both groups of carbonate-bearing plankton appeared during the Mesozoic. The coccolithophorids evolved in the Triassic but began to appear in geologically important concentrations during the Cretaceous and after. By late Cretaceous times they were so abundant that they produced a characteristic rock type: chalk. Similarly, the planktonic foraminifera appeared in great numbers in the Cretaceous.

Foraminifera do not thrive in very-low-oxygen conditions, and the great rise in planktonic foram abundance in the Cretaceous must have been abetted by the rise in oxygen that characterized the latter two periods of the Mesozoic.

## THE END OF THE ERA

The long dinosaur summer probably seemed like an endless summer to the mammals—or might have, if only they had brains large enough to be reflective with. But as most were the size of rats, the Mesozoic mammals probably spent most of their time figuring out how to dodge predators and still get the next meal. Unlike most summers, the Summer of Dinosaurs did not gradually wind down into a cooler autumn. Instead, it turned from summer to the depth and death of winter almost instantaneously. How long was the Chicxulub asteroid in an Earth-crossing orbit before it had its rendezvous with our planet, 65 million years ago? In Chapter 10 we will briefly take a new view of trends in the successors of the dinosaurs—mammals.

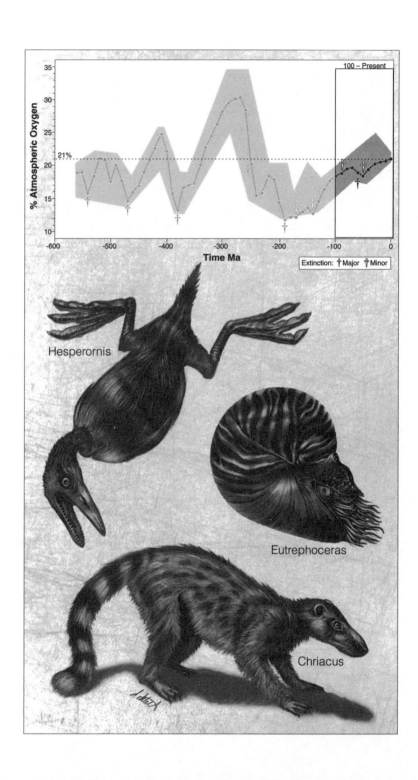

# 10

## THE CRETACEOUS EXTINCTION AND THE
## RISE OF LARGE MAMMALS

So much has been written about the extinction of the dinosaurs that finding anything new to say is a tall task. By mid-2006, the understanding has remained virtually unchanged for a decade or more: a large asteroid smashed the Cretaceous world out of existence in a short-term reign of fire and toxicity. Even weeks after the impact, the damage had been done. The dinosaurs were all dead, as were all ammonites, rudistid clams, mosasaurs, and about 50 percent of the rest of the species that had been blissfully living in the healthy and robust late Cretaceous world.

The dinosaurs disappeared from a world warm and far more oxygenated than it had been when the Mesozoic Era had begun. By the end of the Cretaceous, the world, in terms of its vegetation and atmospheric makeup, had taken on a much more modern cast. But it was not yet the atmosphere of our world. Carbon dioxide was still immensely higher in concentration than now, and—more relevant to the story here—atmospheric oxygen was still lower, although not by much. But the biggest similarity between those times, some 65 million years ago soon after the asteroid impact, and the world of today was that dinosaurs existed in neither. It was the disappearance of the dinosaurs that allowed mammals to fill the emptied world—and fill it they did. Yet here, too, the story of changing oxygen provides a new *why* for some of the most interesting details of that history of "mammalization" of the planet. This chapter looks at how the explosion of mammals filling the new, dinosaur-free world was itself shaped by the last significant rise in oxygen.

## THE HISTORY OF CENOZOIC MAMMALS

As we have seen, the ancestors of the mammals, the Paleozoic and early Mesozoic therapsids, were the dominant land animals until the Permian extinction, and even through the Triassic they maintained moderate diversity, if no longer being the most diverse and numerous of land animals as their Paleozoic forbearers had been. The first true mammals are found in rocks of late Triassic age—at about the time of the first dinosaurs, in fact. But while they may have both appeared at the same time, these two groups then went on to very different fates. The first dinosaurs were already relatively large animals for their time, about a meter long, but soon after they evolved into much larger sizes even before the end of the Triassic. But the first mammals may have been a tenth of the size or less of the first, meter-long dinosaurs, and they then stayed small—for a very long time. *Why?* This obvious question is all the more perplexing because some of the immediate ancestors of the first mammals, even forms that lived alongside them, such as the advanced cynodonts of the late Triassic, were the same size or larger than the early dinosaurs. The first mammals could have been larger, but they were not. Does this size limitation tell us something about their metabolism and constraints in the teeth of the low-oxygen high-heat interval?

Let's recount the ecological position of the first true mammals. It is the late Triassic and early Jurassic. The world is hot. And the world has oxygen levels as low as 10 percent and certainly less than 15 percent for tens of millions of years. Mammals then, as now, are presumed to be warm-blooded and furry. The biggest mystery is how they reproduced. Almost all mammals today are placental, where development takes place in the female to the point that the newborn is capable of existence outside the mother; a few are marsupial, where early birth prior to full development requires a period of time in a secondary pouch within the mother; and a very few, such as the platypus and echidna, lay eggs. Unfortunately, we have no record of mammal eggs, and the fossil record is utterly opaque in telling when the first placental mammals occurred. But there is indirect evidence that is telling. DNA work (based on the molecular clock) indicates that the divergence of major groups of placental mammals—such as divergence into insectivores, carnivores, and artiodactyls, among many other major groups, happened between 100 million and 60 million years ago. Thus, some of the major divisions of mammals predated the Cretaceous extinction. But prior to that extinction all of these forms were small, and not just in the late Cretaceous. All Mesozoic mammals were small, from the first in the Triassic until those of the latest Cretaceous.

One possibility for this small size was so as not to compete with the dinosaurs. No dinosaurs occupied the rodent niche in the Mesozoic, and, as we know all too well today, there is a good living to be had if one is rat-sized or smaller. Under this scenario, mammals did not compete with dinosaurs for ecological reasons. But it may be that other reasons were involved. Perhaps in the hot, low-oxygen world (at least of the late Triassic until the end of the middle Jurassic), a larger, warm-blooded, highly active mammal—an animal that needed to eat much more than a cold-blooded form—was just too energetically expensive to exist. Here, as in Chapter 9, is the idea that dinosaurs were a really different kind of beast than anything we know today and, at least until the Cretaceous, were the only kind of animal that worked really well in the peculiar early and mid-Mesozoic conditions on Earth.

## OXYGEN AND THE SIZES OF MAMMALS

We know that the disappearance of the dinosaurs unleashed a torrent of evolution, producing, in short order, many mammalian taxa. And for the first time, true mammals of larger size evolved. Thus, after the Cretaceous extinction, after the smoke had cleared and the dinosaurs were no longer around, large mammals did begin to appear. But it took a while. Santa Barbara paleontologist John Alroy has meticulously studied the sizes of mammals through time. Through dint of hard work in the library and among numerous dusty museum drawers, he tabulated average size for over 2,000 kinds of mammals from the late Mesozoic and Cenozoic of North America. His work showed an increase in body size through the Cenozoic. But details of the overall size increase indicate that it happened at two different times and rates, rather than as a smooth and continuous increase. The first size increase occurred in the first few million years after the Cretaceous extinction and seems to have been a response by mammals to filling in now-emptied ecological niches. When the herbivorous dinosaurs died out, there was nothing around to browse higher bushes and trees and it became advantageous to grow larger. Large size also lends protection against predation and with the loss of the large, medium, and small dinosaur carnivores, a host of mammal groups began to enlarge. But a much larger increase in size happened much later, in an interval of 50 million to 40 million years ago in the Eocene Epoch. What caused this kind of size increase? One possibility is that it was at least enabled by a rise in oxygen.

The Berner oxygen curve suggests there was a rapid increase in atmospheric oxygen soon after the Cretaceous mass extinction of 65 million years ago. At the same time, mammal size increased. Coincidence? Probably not. As for dinosaurs, it may have been that rising oxygen had something to do with mammalian reproduction. A 2005 study by geochemist Paul Falkowsky suggested that the first appearance of "placental mammals" (the vast majority of mammals today, all of which have live births and nurse their young) could not take place until a critical level of rising oxygen was reached. Their argument was that prior to that time, there was insufficient oxygen within the placenta of pregnant female mammals to nurture developing embryos. This happened in the late Cretaceous.

A key aspect to understanding the evolution of mammals is discovering when the placental form of reproduction first occurred, for there may be a crucial and potentially limiting aspect to placental reproduction related to oxygen. The mother's arterial blood (which is oxygenated) mixes with venous blood (which is enriched in carbon dioxide) in the placenta. Fetal blood picks up its oxygen load from this admixture, and thus the fetus is exposed to oxygen levels that are lower than that of the mother's arterial blood. If the oxygen level of the mother's blood is already depressed because of living at higher altitude or, in the Mesozoic, living at lower atmospheric oxygen levels, the life of the fetus would be endangered.

Today our best model for understanding the life of animals in the lower-oxygen past comes from the study of animals at high altitude. Mammals can readily survive up to 28,000 feet; humans can certainly live, if only for a short time, atop Mt. Everest. But no mammal reproduces above 14,000 feet, which corresponds to the oxygen levels of the early Jurassic. And these are animals that have had 65 million to 100 million years to refine the placental system. The first evolved placentas would surely have been less efficient in delivering oxygen. This would seemingly suggest that the placental system of reproduction was not possible until oxygen blood levels had risen to, perhaps, Cretaceous levels—above 15 percent and perhaps approaching 20 percent. It thus looks as if the oxygen-level increase of the late Cretaceous—an oxygenation event that at least in part helped spark the major diversity increase in dinosaurs—also allowed the first evolution and successful implementation of a new kind of reproductive pattern, placental development (itself but one kind of live birth).

## OXYGEN AND THE RISE OF HUMANS

The change in oxygen levels over time seems to have provoked major changes in evolution. What about one of the biggest of all changes and the most important to us? Can any aspect of the changing atmosphere be interpreted in a new way so as to explain the evolution of our own species? In one sense it can. Johns Hopkins paleontologist Steve Stanley has dubbed us "Children of the Ice Ages," and many paleoanthropologists think that the rise of high intelligence and culture was a means of dealing with the challenges imposed by the glaciation of the past 2 million years, the recent ice ages. But what about oxygen levels? That angle has never been examined.

Almost all models of the atmosphere for the past 10 million years indicate that oxygen levels were higher than now, with atmospheric oxygen levels as high as 28 percent even 5 million years ago. Could it be that higher oxygen levels stimulated, or made easier, the transition to larger brains? Nervous system tissue needs more oxygen than any other kind of tissue, and the rapid enlargement of brain volume perhaps outstripped the circulation system in the brain necessary to sustain all the new neurons. Higher oxygen would give some cushion for error, it seems to me. But this is still sheer speculation.

### THE END OF HISTORY?

We have now finished our journey through time. The 540-million-year trip is over with this chapter. We have seen evidence of major changes in oxygen levels through time. Will oxygen levels continue to change in the future? That is the subject of the final chapter.

## 11

# SHOULD WE FEAR THE OXYGEN FUTURE?

W e have come to the end of history, if history can be counted as something that has already happened. In this last chapter, let's gaze into the future. In keeping with the rest of this book, we will simply ask: can oxygen levels be expected to stay the same, or will they undergo wild swings as they have for the past 540 million years?

The future stretches before us not as one long dark tunnel but as a series of vignettes of variable clarity, like a long avenue punctuated by streetlights of differing luminosity. This century and at least the next will continue to be a time of warming from greenhouse gases produced by humanity. It would be nice if the volcanoes of our world would politely stop outgassing carbon dioxide as well, at least until we humans get our act together and curtail our carbon dioxide production. But the volcanoes just keep spewing this gas into the air, as they have since Earth began. It is our addition to that natural input that is the problem.

We are warming our world, rapidly. A hundred years from now the planet will have returned to its atmospheric condition during the Late Cretaceous through Eocene. Happily, those were times when oxygen was at about its present level, or was even slightly higher. But can we foresee a time farther in the future, when oxygen levels might change? If they do, will they be higher or lower? The fate of oxygen is fixed by the rates at which organic compounds and sulfur-containing compounds and minerals (such as pyrite) are buried, or not, and the rate at which they are weathered or not. Many things control burial and weathering rates. One of these is continental position, and it is this that can best be predicted for the future.

## PLATE TECTONICS AND THE FATE OF GEOGRAPHY

We have been in an ice age now for more than 2.5 million years, and there is a real possibility, based on past experience, that our planet awaits at least that much time again in the seemingly endless cycles of ice and warming for another 2.5 million, or 5 million, or even 10 million years. But eventually the ice ages will end, and indeed with humans producing greenhouse gases at such a prodigious rate it may well be that the time of ice is over, perhaps on timescales of geological periods (tens of millions of years). While global temperature is a function of many factors (including the presence of eroding high mountains, such as the Himalayas, which serve to increase the rate at which the global thermostat removes carbon dioxide from the atmosphere), geography also plays a huge role. Our planet is heading for greater warmth caused both by the increased energy of the sun and by a new global geography. Continental drift is moving enough of the northern landmasses out of high latitudes and moving Antarctica out of the extreme southern latitudes, to end the tyranny of ice. We have a good idea of how and even when that will come about, based on powerful new computer simulations of plate movements.

Plate movements for the past 600 million years are fairly well established (although ambiguity increases for older time). Certainly the positions of continents for the past 200 million years are very well known. Five hundred million years ago, at the time of the major animal diversification known as the Cambrian Explosion, the continents were widely dispersed along the equator. For the next 200 million years, large-scale drift and continental collision resulted in the formation of ever-larger land bodies and major mountain chains, including the Appalachians of the eastern United States. By about 300 million years ago the major continents had coalesced into a single united block, a supercontinent named Pangea. By about 200 million years ago, the huge expanse of land began to break apart and drift in separate ways, creating the Atlantic Ocean as North America split away from Europe and South America from Africa. By 120 million years ago, the southern continents broke apart as well, with Africa, Antarctica, India, and Australia moving in divergent directions and culminating in the continental positions of the modern world.

Given the large number of years that Earth has existed, it is no surprise that the continents have wandered extensively through time. We believe this wandering will not end any time soon, and thus we can confidently predict that continental drift will continue. Furthermore, there is enough information from present-day drifting to allow prediction of future motions. Future motion will have enormous effects on future climate and on the fate of future life.

The best scenarios for understanding future continental change come from labs and analyses that have studied and modeled past motions. We fully expect future directions and rates of plate movements as measured in the present-day to continue into the future. While geographic reconstructions become more problematical and error prone the farther into the future we look, there is good agreement for at least the next 250 million years among the several independent groups of investigators who have examined this problem. The most detailed examination of future continental positions comes from C. Scotese and his Paleomap Project. For nearly two decades Scotese and his coworkers have been compiling maps of continental position in the past and have recently turned their attention to the future. They have arrived at 10 reconstructions of plate positions for the next 250 million years. They are convinced that at the end of this vast period of time there will again be a supercontinent, a return to the state last experienced by Earth at the end of the Paleozoic Era some 250 million years ago.

## A WORLDWIDE BLACK SEA

For the next few million years, plate motions should continue in the current directions. The Atlantic Ocean will continue to widen, and the Pacific Ocean will continue to close. But then we can expect major changes. By 50 million years from now a world map would show fantastic differences from the present-day, and these continental positions are more completely described in my 2003 book coauthored with Don Brownlee, *The Life and Death of Planet Earth*. Perhaps most noticeable would be the loss of the Mediterranean Sea, with its space taken up by an enormous mountain range extending from what is now Europe to the Persian Gulf region. Australia will have moved northward, closing the regions that are now composed of Papua New Guinea and Indonesia, while Baja California will have slid northward along the Pacific Coast of North America.

Far more important than these new continental positions will be the formation of new subduction zones, the regions where Earth's crust dives beneath the continents. Today we know that subduction is initiating in the central Indian Ocean and in the ocean off Puerto Rico. These events suggest that new subduction zones will be in place off both eastern North and South America. As this happens, mountain building will be initiated once again in the Appalachian regions and along the eastern coastline of South America as well. These regions will become home to gigantic active volcanoes and rising mountain chains.

It is not just the positions of the mountains that will change. As Antarctica drifts northwards, its vast ice sheets will melt and the level of the sea will rise. As the sun continues to increase its energy output, it will also cause temperatures to rise, melting other continental ice sheets, with Greenland's being the most important. When both Greenland and Antarctica have seen all of their ice cover melt, the oceans will rise to a sea level nearly 300 feet higher than it is today. Tectonic forces, including the formation of new mid-ocean spreading centers, will also exacerbate sea-level rise. As these form, they will cause the oceans to spill out of their basins and onto low land surfaces.

Flooding of the continental margins brought about by the rise of the sea will cause our planet to undergo a radical climate change. Will it cause there to be a change in oxygen levels as well? As we saw in previous chapters, the two most radical changes in oxygen spanned the interval from the Carboniferous to the end of the Triassic. From the start of the Carboniferous oxygen levels rose to a maximum of over 30 percent by the early Permian. They then began to drop, reaching minimal values of 10 percent some 210 million years ago.

Why was there this one-two punch, and could it be repeated? We know that the interval of time examined included major environmental and biological changes, both of which affected oxygen. The most notable of the former was the coalescing of the continents into Pangea. This amalgamation was finished about the time that oxygen reached its maximal values. Then the continents broke apart at the same time that oxygen levels dropped. Could it simply be that merging continents raise oxygen levels and splitting continents draw down oxygen? That answer may be partially correct. We know why oxygen went up for the first time on land—plants grew into great trees that filled out into planet-spanning forests. As the trees toppled, and topple they did with great regularity, since roots had not yet modernized and all plants of the Carboniferous were less securely rooted, large quantities of organic material were quickly buried before decomposing. In our world, microbes quickly attack felled trees, and the process of decomposition uses oxygen. But 300 million years ago, lignin, the hard parts of trees, was only newly evolved, and microbes of the time had not yet evolved the trick of breaking down this new material. The consequence was the Coal Age. It was the burial of all this organic matter before it could oxidize that caused the rise in temperature, not the merging of continents.

What about the drop in oxygen? Here there may have been more of a connection to continental position. The fused continent caused aridity, and the breakup may have instigated the formation of the Siberian Traps—the biggest flood basalts in Earth's history. The latter caused immense volumes of carbon dioxide to enter the atmosphere, making an already hot planet even hotter. The deep oceans became devoid of oxygen, while on land the rate of forest burial and even the rate of forest growth, dropped markedly. As a consequence, so did oxygen. We must conclude that the breakup of a supercontinent is not good for our planet's biology, as it seems to cause a drop in oxygen.

Can we foresee the formation and then split up of another supercontinent in the far future? The answer seems to be *yes*.

The pattern of smaller continents assembling into a large land mass, or supercontinent, and then breaking apart again over hundreds of million of years of Earth history has been dubbed the Wilson Cycle, in honor of one of the pioneering discoverers of plate tectonics, J. Tuzo

Wilson. The entire cycle seems to take about 500 million years, and there is no reason to believe that plate movement in the near (and far) future will alter this trend. The last supercontinent formation, Pangea, occurred some 300 million years ago, allowing a rough prediction that the next will occur in about 200 million years.

By about 100 million years from now, the continents will have reached their maximum separation and begin to coalesce. By 150 million years from now, the Atlantic will have become far smaller. Subduction will continue along the entire eastern seaboard of both North and South America as the Atlantic Ocean floor is subducted beneath the coastline along gigantic, linear subduction zones. This process will, in turn, create a series of high mountains along the eastern coastlines of those two continents. The Pacific Ocean will increase in size as the continents all begin a mad rush at mutual collision.

By 250 million years into the future, the process of continental amalgamation will have been completed. Europe, Africa, North and South America, and Asia will have formed the supercontinent; however, Antarctica and Australia will not be amalgamated. A curious aspect of this projection is the existence of a large, central equatorial sea. Because of the presence of subduction zones virtually encircling this giant supercontinent, a wall of mountains will enclose much of the land surface, walling off the interiors of the continent, just as the Andes, Sierra Nevada, Cascade, and Coast Range mountains—all high mountains with active volcanism—wall off the interiors of the western parts of North and South America.

But like the last time, this supercontinent will not last. When it breaks apart, it may be that Earth will once again experience a major drop in oxygen, perhaps the equivalent of the Permian mass extinction.

## FUTURE WORK

It will be up to scientists to see how many of the new hypotheses offered in this radical revision of Earth's history are accepted. If even a few are ultimately accepted, it will mean that we will have to revise our understanding of the *whys* in the history of life. If oxygen has varied through time along the lines that Robert Berner and others suggest, it seems highly likely that organisms would adapt in varied ways to these different conditions.

How to test these various ideas? John VandenBrooks—a Ph.D. student of Robert Berner at Yale, as I write this in 2006—is carefully stealing American alligator eggs and raising them in higher-oxygen conditions. He is going to start rearing them in lower temperature and begin the same sort of experiments with parchment eggs, to try to answer questions about why and when these two different kinds of reproduction came about and were subsequently used. Equally important will be studies that measure the rates of oxygen uptake by invertebrates and vertebrates with different kinds of gills. Most important of all, though, is that new generations of scientists spread out across the world, exhuming from the fossil record new clues to better understand the history of life.

Oxygen is to be loved and hated—but also respected. Respiration has been the most important driver of evolution. Will it be an epitaph for our planet?

# REFERENCES

References are listed below as topics that roughly correlate to the various chapters of the book.

## OXYGEN THROUGH TIME (CHAPTERS 1 AND 2)

Baret P, Fouarge A, Bullens P, Lints F. 1994. Life-span of *Drosophila melanogaster* in highly oxygenated atmospheres. *Mech. Ageing Dev.* 76:25–31.

Beerling DJ, Berner RA. 2000. Impact of a Permo-Carboniferous high $O_2$ event on the terrestrial carbon cycle. *Proc. Natl. Acad. Sci. USA* 97:12428–12432.

Beerling DJ, Woodward FI. 2001. *Vegetation and the Terrestrial Carbon Cycle: Modelling the First 400 Million Years.* Cambridge, MA: Cambridge University Press.

Beerling DJ, Woodward FI, Lomas MR, Wills MA, Quick WP, Valdes PJ. 1998. The influence of Carboniferous palaeoatmospheres on plant function: an experimental and modeling assessment. *Phil. Trans. R. Soc. London B* 353:131–140.

Beerling DJ, Lake JA, Berner RA, Hickey LJ, Taylor DW, Royer DL. 2002. Carbon isotope evidence implying high $O_2/CO_2$ ratios in the Permo-Carboniferous atmosphere. *Geochim. Cosmochim. Acta* 66:3757–3767.

Berner RA. 1987. Models for carbon and sulfur cycles and atmospheric oxygen: application to Paleozoic geologic history. *Am. J. Sci.* 287:177–190.

Berner RA. 1999. A new look at the long-term carbon cycle. *GSA Today* 9(11):1–6.

Berner RA. 2001. Modeling atmospheric $O_2$ over Phanerozoic time. *Geochim. Cosmochim. Acta* 65:685–694.

Berner RA. 2002. Examination of hypotheses for the Permo-Triassic boundary extinction by carbon cycle modeling. *Proc. Natl. Acad. Sci. USA* 99:4172–4177.

Berner RA. 2004. *The Phanerozoic Carbon Cycle.* Oxford, UK: Oxford University Press.

Berner RA. 2005. The carbon and sulfur cycles and atmospheric oxygen from middle Permian to middle Triassic. *Geochim. Cosmochim. Acta* 69:3211–3217.

Berner RA. In press. GEOCARBSULF: a combined model for Phanerozoic atmospheric $O_2$ and $CO_2$. *Geochim. Cosmochim. Acta.*

Berner RA, Canfield DE. 1989. A new model of atmospheric oxygen over Phanerozoic time. *Am. J. Sci.* 289:333–361.

Berner RA, Kothavala Z. 2001. GEOCARB III: a revised model of atmospheric $CO_2$ over Phanerozoic time. *Am. J. Sci.* 301:182–204.

Berner RA, Maasch KA. 1996. Chemical weathering and controls on atmospheric $O_2$ and $CO_2$: fundamental principles were enunciated by J.J. Ebelmen in 1845. *Geochim. Cosmochim. Acta* 60:1633–1637.

Berner RA, Raiswell R. 1983. Burial of organic carbon and pyrite sulfur in sediments over Phanerozoic time: a new theory. *Geochim. Cosmochim. Acta* 47:855–862.

Berner RA, Petsch ST, Lake JA, Beerling DJ, Popp BN, et al. 2000. Isotope fractionation and atmospheric oxygen: implications for Phanerozoic $O_2$ evolution. *Science* 287:1630–1633.

Bouverot P. 1985. *Adaptation to Altitude—Hypoxia in Vertebrates.* Berlin: Springer-Verlag.

Briggs DEG. 1985. Gigantism in Palaeozoic arthropods. *Spec. Pap. Paleontol.* 33:157.

Broecker WS, Peacock S. 1999. Anecologic explanation for the Permo-Triassic carbon and sulfur isotope shifts. *Glob. Biogeochem. Cycles* 13:1167–1172.

Carpenter SJ, Lohmann KC. 1997. Carbon isotope ratios of Phanerozoic marine cements: re-evaluating the global carbon and sulfur systems. *Geochim. Cosmochim. Acta* 61:4831–4846.

Carroll RL. 1988. *Vertebrate Paleontology and Evolution.* New York: W. H. Freeman.

Chaloner WG. 1989. Fossil charcoal as an indicator of paleoatmospheric oxygen level. *J. Geol. Soc. London* 146:171–174.

Chapelle G, Peck LS. 1999. Polar gigantism dictated by oxygen availability. *Nature* 399: 114–115.

Clark JS, Cachier H, Goldammer JG, Stocks B, eds. 1997. *Sediment Records of Biomass Burning and Global Change.* New York: Springer-Verlag.

Cloud P. 1976. Beginnings of biospheric evolution and their biogeochemical consequences. *Paleobiology* 2:351–387.

Colman AS, Holland HD. 2000. The global diagenetic flux of phosphorus from marine sediments to the oceans: redox sensitivity and the control of atmospheric oxygen levels. Pp. 53-75 in *Marine Authigenesis: From Microbial to Global,* vol. 66, C Glenn, J Lucas, L Prevot-Lucas, eds. Tulsa, OK: SEPM Special Publication.

Cope MJ, Chaloner WG. 1980. Fossil charcoals as evidence of past atmospheric composition. *Nature* 283:647–649.

Cornette JL, Lieberman B, Goldstein R. 2002. Documenting a significant relationship between macroevolutionary origination rates and Phanerozoic $CO_2$ levels. *Proc. Natl. Acad. Sci. USA* 99:7832-7835.

Crowley TJ, Yip KJ, Baum SK. 1993. Milankovitch cycles and Carboniferous climate. *Geophys. Res. Lett.* 20:1175–1178.

Deeming J, Burgan R, Cohen JD. 1977. *National Fire Danger Rating System—1978.* USDA Gen. Tech. Rep. INT-39. USDA Forest Service, Ogden, UT.

Des Marais DJ, Strauss H, Summons RE, Hayes JM. 1992. Carbon isotope evidence for the stepwise oxidation of the Proterozoic environment. *Nature* 359:605–609.

Dudley R. 1998. Atmospheric oxygen, giant Paleozoic insects and the evolution of aerial locomotor performance. *J. Exp. Biol.* 201:1043–1050.

Dudley R. 2000. *The Biomechanics of Insect Flight: Form, Function, Evolution.* Princeton, NJ: Princeton University Press.

Dudley R, Chai P. 1996. Animal flight mechanics in physically variable gas mixtures. *J. Exp. Biol.* 199:1881–1885.

Ebelmen JJ. 1845. Sur les produits de la decomposition des especes minerales de la famile des silicates. *Annu. Rev. Mines* 12:627–654.

Edmond JM, Measures C, McDuff RE, Chan LH, Collier R, et al. 1979. Ridge crest hydrothermal activity and the balances of the major and minor elements in the ocean: the Galapagos data. *Earth Planet. Sci. Lett.* 46:1–18.

Ellington CP. 1991. Aerodynamics and the origin of insect flight. *Adv. Insect Physiol.* 23:171–210.

Erwin DH. 1993. *The Great Paleozoic Crisis: Life and Death in the Permian.* New York: Columbia University Press.

Falcon-Lang HJ. 2000. Fire ecology of the Carboniferous tropical zone. *Palaeogeogr. Palaeoclimatol. Palaeoecol.* 164:355–371.

Falkowski P. 1997. Evolution of the nitrogen cycle and its influence on the biological sequestration of $CO_2$ in the ocean. *Nature* 387:272–275.

Farquhar GD, Wong SC. 1984. An empirical model of stomatal conductance. *Aust. J. Plant Physiol.* 11:191–210.

Farquhar GD, O'Leary MH, Berry JA. 1982. On the relationship between carbon isotope discrimination and the intercellular carbon dioxide concentration in leaves. *Aust. J. Plant Physiol.* 9:121–137.

Fisher DC. 1979. Evidence for subaerial activity of *Euproops danae* (Merostomata, Xiphosurida). Pp. 379–447 in *Mazon Creek Fossils*, MH Nitecki, ed. New York: Academic Press.

Flower JW. 1964. On the origin of flight in insects. *J. Insect Physiol.* 10:81–88.

Francois LM, Gerard JC. 1986. A numerical model of the evolution of ocean sulfate and sedimentary sulfur during the past 800 million years. *Geochim. Cosmochim. Acta* 50:2289–2302.

Frazier MR, Woods HA, Harrison JF. 2001. Interactive effects of rearing temperature and oxygen on the development of *Drosophila melanogaster*. *Physiol. Biochem. Zool.* 74:641–650.

Gaillardet J, Dupre B, Louvat P, Allegre C. 1999. Global silicate weathering and $CO_2$ consumption rates deduced from the chemistry of large rivers. *Chem. Geol.* 159: 596–611.

Gale J, Rachmilevitch S, Reuveni J, Volokita M. 2001. The high oxygen atmosphere toward the end-Cretaceous: a possible contributing factor to the K/T boundary extinctions and the emergence of C4 species. *J. Exp. Bot.* 52:801–809.

Gans C. 1970. Strategy and sequence in the evolution of the external gas exchangers of ectothermal vertebrates. *Forma Functio.* 3:61–104.

Gans C, Dudley R, Aguilar NM, Graham JB. 1999. Late Paleozoic atmospheres and biotic evolution. *Hist. Biol.* 13:199–219.

Garrels RM, Lerman A. 1984. Coupling of the sedimentary sulfur and carbon cycles—an improved model. *Am. J. Sci.* 284:989–1007.

Garrels RM, Perry EA. 1974. Cycling of carbon, sulfur and oxygen through geologic time. Pp. 303–316 in *The Sea*, vol. 5, E Goldberg, ed. New York: Wiley.

Gilbert DL. 1996. Evolutionary aspects of atmospheric oxygen and organisms. Pp. 1059–1094 in *Environmental Physiology*, vol. 2, MJ Fregly, CM Blatteis, eds. New York: Oxford University Press.

Glasspool I. 2000. A major fire event recorded in the mesofossils and petrology of the Late Permian, Lower Whybrow coal seam, Sydney Basin, Australia. *Palaeogeogr. Palaeoclimatol. Palaeoecol.* 164:373–396.

Graham JB. 1990. Ecological, evolutionary, and physical factors influencing aquatic animal respiration. *Am. Zool.* 30:137–146.

Graham JB, Dudley R, Aguilar N, Gans C. 1995. Implications of the late Palaeozoic oxygen pulse for physiology and evolution. *Nature* 375:117–120.

Graham JB, Aguilar N, Dudley R, Gans C. 1997. The late Paleozoic atmosphere and the ecological and evolutionary physiology of tetrapods. Pp. 141–167 in *Amniote Origins: Completing the Transition to Land*, SS Sumida, KLM Martin, eds. New York: Academic Press.

Greenberg S, Ar A. 1996. Effects of chronic hypoxia, normoxia and hyperoxia on larval development in the beetle *Tenebrio molitor*. *J. Insect Physiol.* 42:991–996.

Greenlee KJ, Harrison JF. 1998. Acid-base and respiratory responses to hypoxia in the grasshopper *Schistorcerca americana*. *J. Exp. Biol.* 201:2843–2855.

Hallam A. 1991. Why was there a delayed radiation after the end-Paleozoic extinctions? *Hist. Biol.* 5:257–262.

Hansen KW, Wallmann K. 2003. Cretaceous and Cenozoic evolution of seawater composition, atmospheric $O_2$ and $CO_2$. *Am. J. Sci.* 303:94–148.

Harrison JF, Lighton JRB. 1998. Oxygen-sensitive flight metabolism in the dragonfly *Erythemis simplicicollis*. *J. Exp. Biol.* 201:1739–1744.

Hayes JM, Strauss H, Kaufman AJ. 1999. The abundance of $^{13}C$ in marine organic matter and isotope fractionation in the global biogeochemical cycle of carbon during the past 800 Ma. *Chem. Geol.* 161:103–125.

Holland HD. 1978. *The Chemistry of the Atmosphere and Oceans*. New York: Wiley Interscience.

Holland HD. 1984. *The Chemical Evolution of the Atmosphere and Oceans*. Princeton, NJ: Princeton University Press.

Hollander DJ, MacKenzie JA. 1991. $CO_2$ control on the carbon-isotope fractionation during aqueous photosynthesis: a paleo-$CO_2$ barometer. *Geology* 19:929–932.

Holter P, Spangenberg A. 1997. Oxygen uptake in coprophilous beetles (Aphodius, Geotrupes, Sphaeridium) at low oxygen and high carbon dioxide concentrations. *Physiol. Entomol.* 22:339–343.

Jin Y, Wang Y, Wang W, Shang QH, Cao CQ, Erwin D. 2000. Pattern of marine mass extinction near the Permian-Triassic boundary in south China. *Science* 289:432–436.

Jones TP, Chaloner WG. 1991. Fossil charcoal, its recognition and palaeoatmospheric significance. *Palaeogeogr. Palaeoclimatol. Palaeoecol.* 97:39–50.

Kadko D. 1996. Radioisotopic studies of submarine hydrothermal vents. *Rev. Geophys.* 34:349–366.

Kim S-T, O'Neil JR. 1997. Equilibrium and nonequilibrium oxygen isotope effects in synthetic carbonates. *Geochim. Cosmochim. Acta* 61:3461–3475.

Kingsolver JG, Koehl MAR. 1994. Selective factors in the evolution of insect wings. *Annu. Rev. Entomol.* 39:425–451.

Kloek G, Ridgel G, Ralin D. 1976. Survivorship and life expectancy of *Drosophila melanogaster* populations in abnormal oxygen-normal pressure regimes. *Aviat. Space Environ. Med.* 47:1174–1176.

Knoll AH, Bambach RK, Canfield DE, Grotzinger JP. 1996. Comparative earth history and late Permian mass extinction. *Science* 273:452–457.

Komrek EV. 1973. Ancient fires. Pp. 219–240 in *Proceedings of the Annual Tall Timbers Fire Ecology Conference, June 8–9, Lubbock, Texas.* Tallahassee, FL: Tall Timbers Research Station.

Kukalová-Peck J. 1985. Ephemeroid wing venation based upon new gigantic Carboniferous mayflies and basic morphology, phylogeny, and metamorphosis of pterygote insects (Insecta, Ephemerida). *Can. J. Zool.* 63:933–955.

Kukalová-Peck J. 1987. New Carboniferous Diplura, Monura, and Thysanura, the hexapod ground plan, and the role of thoracic lobes in the origin of wings (Insecta). *Can. J. Zool.* 65:2327–2345.

Kump LR. 1988. Terrestrial feedback in atmospheric oxygen regulation by fire and phosphorus. *Nature* 335:152–154.

Kump LR. 1989. Alternative modeling approaches to the geochemical cycles of carbon, sulfur and strontium isotopes. *Am. J. Sci.* 289:390–410.

Kump LR. 1989. Chemical stability of the atmosphere and ocean. *Glob. Planet. Change* 1:123–136.

Kump LR, Arthur MA. 1999. Interpreting carbon-isotope excursions: carbonates and organic matter. *Chem. Geol.* 161:181–198.

Kump LR, Garrels RM. 1986. Modeling atmospheric $O_2$ in the global sedimentary redox cycle. *Am. J. Sci.* 286:336–360.

Labandeira CC. 1998. Early history of arthropod and vascular plant associations. *Annu. Rev. Earth Planet. Sci.* 26:329–377.

Lasaga AC. 1989. A new approach to the isotopic modeling of the variation of atmospheric oxygen through the Phanerozoic. *Am. J. Sci.* 298:411–435.

Lasaga AC, Ohmoto H. 2002. The oxygen geochemical cycle: dynamics and stability. *Geochim. Cosmochim. Acta* 66:361–381.

Lawlor DW. 2001. *Photosynthesis.* New York: Springer.

Laws EA, Popp BN, Bidigare RR, Kennicutt MC, Macko SA. 1995. Dependence of phytoplankton carbon isotopic composition on growth rate and $[CO_2]$aq: theoretical considerations and experimental results. *Geochim. Cosmochim. Acta* 59:1131–1138.

Lenton TM, Watson AJ. 2000. Redfield revisited 2: what regulates the oxygen content of the atmosphere? *Glob. Biogeochem. Cycles* 14(1):249–268.

Loudon C. 1988. Development of *Tenebrio molitor* in low oxygen levels. *J. Insect Physiol.* 34:97–103.

Loudon C. 1989. Tracheal hypertrophy in mealworms: design and plasticity in oxygen supply systems. *J. Exp. Biol.* 147:217–235.

Lovelock J. 1988. *The Ages of Gaia.* New York: Norton.

Mangum CP. 1997. Invertebrate blood oxygen carriers. Pp. 1097–1135 in *Handbook of Physiology*, vol. 2. New York: Oxford University Press.

Mii H, Grossman EL, Yancey TE. 1999. Carboniferous isotope stratigraphies of North America: implications for Carboniferous paleoceanography and Mississippian glaciation. *Geol. Soc. A. Bull.* 111:960–973.

Mill PJ. 1985. Structure and physiology of the respiratory system. Pp. 517–593 in *Comprehensive Insect Physiology, Biochemistry, and Pharmacology*, vol. 3, GA Kerkut, LI Gilbert, eds. Oxford, UK: Pergamon Press.

Milliman JD, Syvitski JPM. 1992. Geomorphic tectonic control of sediment discharge to the ocean: the importance of small mountainous rivers. *J. Geol.* 100:535–544.

Morton JL, Sleep NH. 1985. A mid-ocean ridge thermal model-constraints on the volume of axial hydrothermal heat-flux. *J. Geophys. Res. Solids* 90:134–153.

Nelson M. 2001. A dynamical systems model of the limiting oxygen index test: II. Retardancy due to char formation and addition of inert fillers. *Combust. Theory Model.* 5:59–83.

Orr WC, Sohal RS. 1994. Extension of lifespan by overexpression of superoxide dismutase and catalase in *Drosophila melanogaster*. *Science* 263:1128–1130.

Petsch ST. 1999. Comment on "Carbon isotope ratios of Phanerozoic marine cements: re-evaluating global carbon and sulfur systems," by SJ Carpenter and KC Lohmann, 1997, *Geochim. Cosmochim. Acta* 61:4831–4846. *Geochim. Cosmochim. Acta* 63:307.

Petsch ST, Berner RA. 1998. Coupling the long-term geochemical cycles of carbon, phosphorus, sulfur, and iron: the effect on atmospheric $O_2$ and the isotopic records of carbon and sulfur. *Am. J. Sci.* 298:246–262.

Petsch ST, Berner RA, Eglinton TI. 2000. A field study of the chemical weathering of ancient sedimentary organic matter. *Org. Geochem.* 31:475–487.

Rachmilevitch S, Reuveni J, Pearcy RW, Gale J. 1999. A high level of atmospheric oxygen, as occurred toward the end Cretaceous period, increased leaf diffusion conductance. *J. Exp. Bot.* 50:869–872.

Rashbash D, Langford B. 1968. Burning of wood in atmospheres of reduced oxygen concentration. *Combust. Flame* 12:33–40.

Raven JA. 1991. Plant responses to high $O_2$ concentrations: relevance to previous high $O_2$ episodes. *Palaeogeogr. Palaeoclimatol. Palaeoecol.* 97:19–38.

Raven JA, Johnston AM, Parsons R, Kubler J. 1994. The influence of natural and experimental high $O_2$ concentrations on $O_2$-evolving phototrophs. *Biol. Rev.* 69:61–94.

Rayner JMV. 2002. Palaeogeophysics and the limits to design: gravity, the atmosphere, and the evolution of animal locomotion. In *Planet Earth*, A Lister, L Rothschild, eds. London: Academic Press.

Reid RC, Prausnitz JM, Poling BE. 1987. *The Properties of Gases and Liquids*, 4th ed. New York: McGraw-Hill.

Robinson JM. 1989. Phanerozoic $O_2$ variation, fire, and terrestrial ecology. *Palaeogeogr. Palaeoclimatol. Palaeoecol.* 75:223–240.

Robinson JM. 1991. Phanerozoic atmospheric reconstructions: a terrestrial perspective. *Global Planet. Changes* 5:51–62.

Rogers AD. 2000. The role of the oceanic oxygen minima in generating biodiversity in the deep sea. *Deep-Sea Res.* 47:119–148.

Ronov AB. 1976. Global carbon geochemistry, volcanism, carbonate accumulation and life. *Geochem. Int.* 13:172–195.

Rowley DB. 2002. Rate of plate creation and destruction: 180 Ma to present. *Geol. Soc. Am. Bull.* 114:927–933.

Runnegar B. 1982. The Cambrian explosion: animals or fossils? *J. Geol. Soc. Aust.* 29: 395–411.

Runnegar B. 1982. Oxygen requirements, biology and phylogenetic significance of the late Precambrian worm *Dickinsonia*, and the evolution of the burrowing habit. *Alcheringa* 6:223–239.

Sanudo-Wilhelmy SA, Kustka AB, Gobler CJ, Hutchins DA, Yang M, et al. 2001. Phosphorus limitation of nitrogen fixation by Trichodesmium in the central Atlantic Ocean. *Nature* 411:66–69.

Scott AC. 2000. The pre-Quaternary history of fire. *Palaeogeogr. Palaeoclimatol. Palaeoecol.* 164:335–371.

Shear W. 1991. The early development of terrestrial ecosystems. *Science* 351:283–289.

Shear WA, Kukalová-Peck J. 1990. The ecology of Paleozoic terrestrial arthropods: the fossil evidence. *Can. J. Zool.* 68:1807–1834.

Shigenaga MK, Hagen TM, Ames BN. 1994. Oxidative damage and mitochondrial decay in aging. *Proc. Natl. Acad. Sci. USA* 91:10771–10778.

Smith RMH, Ward PD. 2001. Pattern of vertebrate extinctions across an event bed at the Permian-Triassic boundary in the Karoo Basin of South Africa. *Geology* 29:1147–1150.

Spicer JI, Gaston KJ. 1999. Amphipod gigantism dictated by oxygen availability? *Ecol. Lett.* 2:397–403.

Stallard RF. 1985. River chemistry, geology, geomorphology, and soils in the Amazon and Orinoco Basins. Pp. 293–316 in *The Chemistry of Weathering*, J Drever, ed. Boston, MA: D. Reidel.

Strauss H. 1999. Geological evolution from isotope proxy signals—sulfur. *Chem. Geol.* 161:89–101.

Sussot RA. 1980. Effect of heating rate on char yield from forest fuels. USDA Forest Service Research Note INT-295. USDA Forest Service, Ogden, UT.

Tewarson A. 2000. Nonmetallic material flammability in oxygen-enriched atmospheres. *J. Fire Sci.* 18:183–214.

Ultsch GR. 1974. Gas exchange in the Sirenidae (Amphibia, Caudata). I. Oxygen consumption of submerged sirenids as a function of body size and respiratory surface area. *Comp. Biochem. Physiol.* 47A:485–498.

Van Cappellen P, Ingall ED. 1996. Redox stabilization of the atmosphere and oceans by phosphorus-limited marine productivity. *Science* 271:493–496.

Veizer J, Holser WT, Wilgus CK. 1980. Correlation of C-13-C-12 and S-34-S-32 secular variations. *Geochim. Cosmochim. Acta* 44:579–587.

Veizer J, Ala D, Azmy K, Bruckschen P, Buhl D, et al. 1999. $^{87}Sr/^{86}Sr$, $\pm^{13}C$ and $\pm^{18}O$ evolution of Phanerozoic seawater. *Chem. Geol.* 161:59–88.

Vogel S. 1994. *Life in Moving Fluids: The Physical Biology of Flow.* Princeton, NJ: Princeton University Press.

Walker JCG. 1986. Global geochemical cycles of atmospheric oxygen. *Marine Geol.* 70: 159–174.

Watson A, Lovelock JE, Margulis L. 1978. Methanogenesis, fires, and the regulation of atmospheric oxygen. *Biosystems* 10:293–298.

Watson AJ. 1978. Consequences for the biosphere of grassland and forest fires. PhD dissertation, Reading University, United Kingdom.

Weis-Fogh T. 1964. Diffusion in insect wing muscle, the most active tissue known. *J. Exp. Biol.* 41:229–256.

Weis-Fogh T. 1964. Functional design of the tracheal system of flying insects as compared with the avian lung. *J. Exp. Biol.* 41:207–227.

Wignall PB, Twitchett RJ. 1996. Oceanic anoxia and the end Permian mass extinction. *Science* 272:1155–1158.

Wootton RJ. 1988. The historical ecology of aquatic insects: an overview. *Palaeogeogr. Palaeoclimatol. Palaeoecol.* 62:477–492.

Wootton RJ. 1990. Major insect radiations. Pp. 187–208 in *Major Evolutionary Radiations*, PD Taylor, GP Larwood, eds. Oxford, UK: Clarendon Press.

Wootton RJ, Ellington CP. 1991. Biomechanics and the origin of insect flight. Pp. 99–112 in *Biomechanics in Evolution*, JMV Rayner, RJ Wootton, eds. Cambridge, MA: Cambridge University Press.

## CAMBRIAN EXPLOSION (CHAPTER 3)

Allison PA, Brett CE. 1995. In situ benthos and paleo-oxygenation in the Middle Cambrian Burgess Shale, British Columbia, Canada. *Geology* 23:1079–1082.

Aronson RB. 1992. Decline of the Burgess Shale fauna: ecologic or taphonomic restriction? *Lethaia* 25:225–229.

Bergström J. 1986. *Opabinia* and *Anomalocaris*, unique Cambrian "arthropods". *Lethaia* 19:241–246.

Bousfield EL. 1995. A contribution to the natural classification of Lower and Middle Cambrian arthropods: food-gathering and feeding mechanisms. *Amphipacifica* 2:3–34.

Briggs DEG. 1978. The morphology, mode of life, and affinities of *Canadaspis perfecta* (Crustacea: Phyllocarida), Middle Cambrian, Burgess Shale, British Columbia. *Phil. Trans. R. Soc. London B* 281:439–487.

Briggs DEG. 1979. *Anomalocaris*, the largest known Cambrian arthropod. *Palaeontology* 22:631–664.

Briggs DEG. 1992. Phylogenetic significance of the Burgess Shale crustacean *Canadaspis*. *Acta Zool. (Stockholm)* 73:293–300.

Briggs DEG, Fortey RA. 1989. The early radiation and relationships of the major arthropod groups. *Science* 246:241–243.

Briggs DEG, Whittington HB. 1985. Modes of life of arthropods from the Burgess Shale, British Columbia. *Phil. Trans. R. Soc. Edinburgh* 76:149–160.

Bruton DL, Jensen A, Jacquet R. 1985. The use of models in the understanding of Cambrian arthropod morphology. *Trans. R. Soc. Edinburgh* 76:365–369.

Budd G. 1993. A Cambrian gilled lobopod from Greenland. *Nature* 364:709–711.

Budd GE. 1996. The morphology of *Opabinia regalis* and the reconstruction of the arthropod stem-group. *Lethaia* 29:1–14.

Butterfield NJ. 1990. Organic preservation of non-mineralizing organisms and the taphonomy of the Burgess Shale. *Paleobiology* 16:272–286.

Butterfield NJ. 1990. A reassessment of the enigmatic Burgess Shale fossil *Wiwaxia corrugata* (Matthew) and its relationship to the polychaete *Canadia spinosa* Walcott. *Paleobiology* 16:287–303.

Butterfield NJ. 1997. Plankton ecology and the Proterozoic-Phanerozoic transition. *Paleobiology* 23:247–262.

Chen J, Ramsköld L, Zhou G. 1994. Evidence for monophyly and arthropod affinity of Cambrian predators. *Science* 264:1304–1308.

Chen J, Zhou G, Ramsköld L. 1995. A new Early Cambrian onychophoran-like animal *Paucipodia* gen. nov., from the Chengjiang fauna, China. *Trans. R. Soc. Edinburgh: Earth Sci.* 85:275–282.

Chen J, Edgecombe GD, Ramsköld L. 1997. Morphological and ecological disparity in naraoiids (Arthropoda) from the Early Cambrian Chengjiang fauna, China. *Rec. Aust. Mus.* 49:1–24.

Collins D. 1996. The "evolution" of *Anomalocaris* and its classification in the arthropods Class Dinocarida (Nov.) and Order Radiodonta (Nov.). *J. Paleontol.* 70:280–293.

Collins D, Briggs D, Conway Morris S. 1983. New Burgess Shale fossil sites reveal Middle Cambrian faunal complex. *Science* 222:163–167.

Conway Morris S. 1977. A new entoproct-like organism from the Burgess Shale of British Columbia. *Palaeontology* 20:833–845.

Conway Morris S. 1977. Fossil priapulid worms. *Spec. Pap. Palaeontol.* 20:i–iv, 1–95.

Conway Morris S. 1979. The Burgess Shale (Middle Cambrian) fauna. *Annu. Rev. Ecol. System.* 10:327–349.

Conway Morris S. 1979. Middle Cambrian polychaetes from the Burgess Shale of British Columbia. *Phil. Trans. R. Soc. London B* 285:227–274.

Conway Morris S, ed. 1982. *Atlas of the Burgess Shale.* London: Palaeontological Association.

Conway Morris S. 1985. The Middle Cambrian metazoan *Wiwaxia corrugata* (Matthew) from the Burgess Shale and *Ogygopsis* Shale, British Columbia, Canada. *Phil. Trans. R. Soc. London* 307:507–586.

Conway Morris S. 1985. Cambrian Lagerstätten: their distribution and significance. *Phil. Trans. R. Soc. London B* 311:49–65.

Conway Morris S. 1986. The community structure of the Middle Cambrian Phyllopod bed (Burgess Shale). *Palaeontology* 29:423–467.

Conway Morris S. 1989. Burgess Shale faunas and the Cambrian explosion. *Science* 246:339–346.

Conway Morris S. 1989. The persistence of Burgess Shale-type faunas: implications for the evolution of deeper-water faunas. *Trans. R. Soc. Edinburgh: Earth Sci.* 80: 271–283.

Conway Morris S. 1990. Late Precambrian and Cambrian soft-bodied faunas. *Annu. Rev. Earth Planet. Sci.* 18:101–122.

Conway Morris S. 1992. Burgess Shale-type faunas in the context of the "Cambrian explosion": a review. *J. Geol. Soc. London* 149:631–636.

Conway Morris S. 1993. Ediacaran-like fossils in Cambrian Burgess Shale-type faunas of North America. *Palaeontology* 36:593–635.

Conway Morris S. 1993. The fossil record and the early evolution of the Metazoa. *Nature* 361:219–225.

Conway Morris S. 1994. Why molecular biology needs palaeontology. *Development* (Suppl.):1–13.

Conway Morris S, Peel JS. 1995. Articulated halkieriids from the Lower Cambrian of North Greenland and their role in early protostome evolution. *Phil. Trans. R. Soc. London B* 347:305–358.

Conway Morris S, Robison RA. 1982. The enigmatic medusoid *Peytoia* and a comparison of some Cambrian biotas. *J. Paleontol.* 56:116–122.

Conway Morris S, Whittington HB. 1985. Fossils of the Burgess Shale, a national treasure in Yoho National Park, British Columbia. *Misc. Rep. Geol. Surv. Can.* 43:1–31.

Degan S, Conway Morris S, Xianliang Z. 1996. A *Pikaia*-like chordate from the Lower Cambrian of China. *Nature* 384:157–158.

Degan S, Xingliang Z, Ling C. 1996. Reinterpretation of *Yunnanozoon* as the earliest known hemichordate. *Nature* 380:428–430.

Dzik J. 1995. *Yunnanozoon* and the ancestry of chordates. *Acta Palaeontol. Pol.* 40:341–360.

Erwin DM. 1993. The origin of metazoan development: a palaeobiological perspective. *Biol. J. Linn. Soc.* 50:255–274.

Fritz WH. 1971. Geological setting of the Burgess Shale. Pp. 1155–1170 in *Symposium on Extraordinary Fossils. Proceedings of the North American Paleontological Convention, Field Museum of Natural History, Chicago, September 5–7, 1969, Part I.* Lawrence, KS: Allen Press.

Gould SJ. 1986. *Wonderful Life.* New York: Norton.

Gould SJ. 1991. The disparity of the Burgess Shale arthropod fauna and the limits of cladistic analysis: why we must strive to quantify morphospace. *Paleobiology* 17: 411–423.

Grotzinger JP, Bowring SA, Saylor BZ, Kaufman AJ. 1995. Biostratigraphic and geochronologic constraints on early animal evolution. *Science* 270:598–604.

Hou X, Bergström J. 1995. Cambrian lobopodians—ancestors of extant onychophorans? *Zool. J. Linn. Soc.* 114:3–19.

Hou X, Ramsköld L, Bergström J. 1991. Composition and preservation of the Chengjiang fauna—a Lower Cambrian soft-bodied biota. *Zool. Scrip.* 20:395–411.

Hou X, Bergström J, Ahlberg P. 1995. *Anomalocaris* and other large animals in the Lower Cambrian of southwest China. *Geol. Fören. Stockh. Förhandl.* 117:163–183.

Huilin L, Shixue H, Shishan Z, Yonghe T. 1997. New occurrence of the early Cambrian Chengjiang fauna from Haikou, Kunming, Yunnan province, and study of Trilobitoidea. *Acta Geol. Sin.* 71:97–104.

Kirschvink JL, Magaritz M, Ripperdan RL, Zhuravlev AY, Rozanov AY. 1991. The Precambrian-Cambrian boundary: magnetostratigraphy and carbon isotopes resolve correlation problems between Siberia, Morocco, and South China. *GSA Today* 1: 69–91.

Kirschvink JL, Ripperdan RL, Evans DA. 1997. Evidence for a large-scale Early Cambrian reorganization of continental masses by inertial interchange true polar wander. *Science* 277:541–545.

Ludvigsen R. 1989. The Burgess Shale: not in the shadow of the cathedral escarpment. *Geosci. Can.* 16:51–59.

Mankiewicz C. 1992. *Obruchevella* and other microfossils in the Burgess Shale: preservation and affinity. *J. Paleontol.* 66:717–729.

McMenamin MAS, McMenamin DLS. 1990. *The Emergence of Animals—The Cambrian Breakthrough.* New York: Columbia University Press.

Nedin C. 1995. The Emu Bay Shale, a Lower Cambrian fossil Lagerstätten, Kangaroo Island, South Australia. *Mem. Assoc. Aust. Palaeontol.* 18:31–40.

Ramsköld L. 1992. Homologies in Cambrian Onychophora. *Lethaia* 25:443–460.

Ramsköld L, Xianguang H. 1991. New early Cambrian animal and onychophoran affinities of enigmatic metazoans. *Nature* 351:225–228.

Rigby JK. 1986. Sponges of the Burgess Shale (Middle Cambrian), British Columbia. *Palaeontogr. Can.* 2:1–105.

Robison RA, Wiley EO. 1995. A new arthropod, *Meristosoma*: more fallout from the Cambrian explosion. *J. Paleontol.* 69:447–459.

Simonetta AM, Conway Morris S, eds. 1991. *The Early Evolution of Metazoa and the Significance of Problematic Taxa.* Cambridge, UK: Cambridge University Press.

Simonetta AM, Insom E. 1993. New animals from the Burgess Shale (Middle Cambrian) and their possible significance for the understanding of the Bilateria. *Boll. Zool.* 60:97–107.

Towe KM. 1996. Fossil preservation in the Burgess Shale. *Lethaia* 29:107–108.

Valentine JW. 1994. Late Precambrian bilaterans: grades and clades. *Proc. Natl. Acad. Sci. USA* 91:6751–6757.

Valentine JW, Erwin DH, Jablonski D. 1996. Developmental evolution of metazoan body plans: the fossil evidence. *Dev. Biol.* 173:373–381.

Whittington HB. 1971. The Burgess Shale: history of research and preservation of fossils. Pp. 1170–1201 in *Symposium on Extraordinary Fossils. Proceedings of the North American Paleontological Convention, Field Museum of Natural History, Chicago, September 5–7, 1969, Part I.* Lawrence, KS: Allen Press.

Whittington HB. 1975. The enigmatic animal *Opabinia regalis*, Middle Cambrian, Burgess Shale, British Columbia. *Phil. Trans. R. Soc. London B* 271:1–43.

Whittington HB. 1975. Trilobites with appendages from the Middle Cambrian, Burgess Shale, British Columbia. *Foss. Strata* 4:97–136.

Whittington HB. 1979. Early arthropods, their appendages and relationships. Pp. 253–268 in *The Origin of Major Invertebrate Groups*, MR House, ed. London: Academic Press.

Whittington HB. 1980. The significance of the fauna of the Burgess Shale, Middle Cambrian, British Columbia. *Proc. Geol. Assoc.* 91:118–127.

Whittington HB. 1981. Rare arthropods from the Burgess Shale, Middle Cambrian, British Columbia. *Phil. Trans. R. Soc. London B* 292:329–357.

Whittington HB, Briggs DEG. 1985. The largest Cambrian animal, *Anomalocaris*, Burgess Shale, British Columbia. *Phil. Trans. R. London B* 309:569–609.

Wills MA, Briggs DEG, Fortey RA. 1994. Disparity as an evolutionary index: a comparison of Cambrian and Recent arthropods. *Paleobiology* 20:93–130.

Yochelson EL. 1996. Discovery, collection, and description of the Middle Cambrian Burgess Shale biota by Charles Doolittle Walcott. *Proc. Am. Phil. Soc.* 140:469–545.

## ANIMAL EVOLUTION AND EXTINCTION

Alvarez L, Alvarez W, Asaro F, Michel H. 1980. Extra-terrestrial cause for the Cretaceous-Tertiary extinction. *Science* 208:1094–1108.

Benton M. 1995. Diversification and extinction in the history of life. *Science* 268:52–58.

Covey C, Thompson S, Weissman P, Maccracen M. 1994. Global climatic effects of atmospheric dust from and asteroid or comet impact on earth. *Glob. Planet. Change* 9:263–273.

Ellis J, Schramm D. 1995. Could a supernova explosion have caused a mass extinction? *Proc. Natl. Acad. Sci. USA* 92:235–238.

Erwin D. 1993. *The Great Paleozoic Crisis: Life and Death in the Permian.* New York: Columbia University Press.

Erwin D. 1994. The Permo-Triassic extinction. *Nature* 367:231–236.

Hallam A. 1994. The earliest Triassic as an anoxic event, and its relationship to the end-Paleozoic mass extinction. *Can. Soc. Petrol. Geol. Mem.* 17:797–804.

Hallam A, Wignall P. 1997. *Mass Extinctions and Their Aftermath.* Oxford, UK: Oxford University Press.

Hsu K, Mckenzie J. 1990. Carbon isotope anomalies at era boundaries: global catastrophes and their ultimate cause. *Geol. Soc. Am. Spec. Pap.* 247:61–70.

Kauffman EG, Walliser OH, eds. 1990. *Extinction Events in Earth History.* New York: Springer-Verlag.

Knoll A, Bambach R, Canfield D, Grotzinger J. 1996. Comparative earth history and Late Permian mass extinction. *Science* 273:452–457.

Marshall C. 1990. Confidence intervals on stratigraphic ranges. *Paleobiology* 16:1–10.

Marshall C, Ward P. 1996. Sudden and gradual molluscan extinctions in the latest Cretaceous of Western European Tethys. *Science* 274:1360–1363.

McRoberts C. 2004. Marine bivalves and the end-Triassic mass extinction: faunal turnover, isotope anomalies and implications for the position of the Triassic/Jurassic. P. 1138 in *32nd International Geological Congress,* Florence, Italy, August 20–28.

Morante R. 1996. Permian and early Triassic isotopic records of carbon and strontium events in Australia and a scenario of events about the Permian-Triassic boundary. *Hist. Geol.* 11:289–310.

Olsen PE, Kent DV, Sues HD, Koeberl C, Huber H, Montanari A, Rainforth EC, Fowell SJ, Szajna MJ, Hartline BW. 2002. Ascent of dinosaurs linked to an irridium anomaly at the Triassic-Jurassic boundary. *Science* 296:1305–1307.

Pope K, Baines A, Ocampo A, Ivanov B. 1994. Impact winter and the Cretaceous Tertiary extinctions: results of a Chicxulub asteroid impact model. *Earth Planet. Sci. Exp.* 128:719–725.

Powell M. 2005. Climatic basis for sluggish macroevolution during the late Paleozoic ice age. *Geology* 33:381–384.

Rampino M, Caldeira K. 1993. Major episodes of geologic change: correlations, time structure and possible causes. *Earth Planet. Sci. Lett.* 114:215–227.

Raup D. 1979. Size of the Permo-Triassic bottleneck and its evolutionary implications. *Science* 206:217–218.

Raup D. 1990. Impact as a general cause of extinction: a feasibility test. In *Global Catastrophes in Earth History*, V Sharpton, P Ward, eds. *Geol. Soc. Am. Spec. Pap.* 247:27–32.

Raup D. 1991. A kill curve for Phanerozoic marine species. *Paleobiology* 17:37–48.

Raup D, Sepkoski J. 1984. Periodicity of extinction in the geologic past. *Proc. Natl. Acad. Sci. USA* 81:801–805.

Retallack G. 1995. Permian-Triassic crisis on land. *Science* 267:77–80.

Schindewolf O. 1963. Neokatastrophismus? *Z. Deutsch. Geol. Gesell.* 114:430–445.

Schultz P. 1994. Impact angle effects on global lethality. Geological Society of multiring impact basin: evaluation of geophysical data, well logs, and drill core samples. *Lunar and Planetary Institute Contribution* 825:108–110.

Sheehan P, Fastovsky D, Hoffman G, Berghaus C, Gabriel D. 1991. Sudden extinction of the dinosaurs: latest Cretaceous, Upper Great Plains, U.S.A. *Science* 254:835–839.

Sigurdsson H, D'hondt S, Carey S. 1992. The impact of the Cretaceous-Tertiary bolide on evaporite terrain and generation of major sulfuric acid aerosol. *Earth Planet. Sci. Lett.* 109:543–559.

Stanley S. 1987. *Extinctions.* New York: W. H. Freeman.

Stanley S, Yang X. 1994. A double mass extinction at the end of the Paleozoic Era. *Science* 266:1340–1344.

Teichert C. 1990. The end-Permian Extinction. Pp. 161–190 in *Global Events in Earth History*, E Kauffman, O Walliser, eds. New York: Springer-Verlag.

Ward P. 1990. The Cretaceous/Tertiary extinctions in the marine realm: a 1990 perspective. *Geol. Soc. Am. Spec. Pap.* 247:425–432.

Ward P. 1990. A review of Maastrichtian ammonite ranges. *Geol. Soc. Am. Spec. Pap.* 247:519–530.

Ward P. 1994. *The End of Evolution.* New York: Bantam Doubleday Dell.

Ward P, Kennedy W. 1993. Maastrichtian ammonites from the Biscay region (France and Spain). *Paleontol. Soc. Memoir* 34:1–58.

Ward P, Kennedy WJ, MacLeod K, Mount J. 1991. Ammonite and inoceramid bivalve extinction patterns in Cretaceous-Tertiary boundary sections of the Biscay Region (southwest France, northern Spain). *Geology* 19:1181.

## PERMIAN EVENTS AND FAUNAS

Anderson J, Cruickshank A. 1978. The biostratigraphy of the Permian and the Triassic: a review of the classification and distribution of Permo-Triassic tetrapods. *Palaentol. Afr.* 21:15–44.

Broom R. 1932. *The Mammal-like Reptiles of South Africa and the Origin of Mammals.* London: Witherby.

Erwin D. 1993. *The Great Paleozoic Crisis: Life and Death in the Permian.* New York: Columbia University Press.

Erwin D. 1994. The Permo-Triassic extinction. *Nature* 367:231–236.

Holser W, et al. 1989. A unique geochemical record at the Permian-Triassic boundary. *Nature* 337:39–44.

Hotton N. 1967. Stratigraphy and sedimentation in the Beaufort Series (Permian-Triassic), South Africa. Pp. 390-428 in *Essays in Paleontology and Stratigraphy*, C Teichert, EL Yochelson, eds. Lawrence, KS: University of Kansas, Department of Geology Special Publication.

Hsu K, Mckenzie J. 1990. Carbon isotope anomalies at era boundaries: global catastrophes and their ultimate cause. *Geol. Soc. Am. Spec. Pap.* 247:61–70.

Isozaki Y. 1994. Superanoxia across the Permo-Triassic boundary: record in accreted deep-sea pelagic chert in Japan. Pp. 805–812 in *Global Environments and Resources*. Canadian Society Petroleum Geologists, Memoir 17.

Jablonski D. 1991. Extinctions: a paleontological perspective. *Science* 253:754–757.

Keyser A, Smith R. 1979. Vertebrate biozonation of the Beaufort Group with special reference to the Western Karoo Basin. *Mem. Geol. Surv. S. Afr.* 12:1–36.

King G. 1990. Dicynodonts and the end Permian event. *Palaentol. Afr.* 27:31–39.

Kirchner JW, Weil A. 2000. Delayed biological recovery from extinctions throughout the fossil record. *Nature* 404:177–180.

Kitching J. 1977. Distribution of the Karoo vertebrate fauna. *Mem. B. Price Inst. Palaentol. Res.* 1:1–131.

Knoll A, Bambach R, Canfield D, Grotzinger J. 1996. Comparative darth history and Late Permian mass extinction. *Science* 273:452–457.

Kruess A, Tscharntke T. 1994. Habitat fragmentation, species loss and biological control. *Science* 264:1581–1584.

Morante R. 1996. Permian and early Triassic isotopic records of carbon and strontium events in Australia and a scenario of events about the Permian-Triassic boundary. *Hist. Geol.* 11:289–310.

Raup D. 1979. Size of the Permo-Triassic bottleneck and its evolutionary implications. *Science* 206:217–218.

Retallack G. 1995. Permian-Triassic crisis on land. *Science* 267:77–80.

Retallack G. 1999. Postapocalyptic greenhouse paleoclimate revealed by Earliest Triassic paleosols in the Sandhy Basin, Australia. *GSA Bull.* 111:52–70.

Retallack G, Krull E. 1999. Landscape ecological shift at the Permian/Triassic boundary in Antarctica. *Aust. J. Earth Sci.* 46:785–812.

Rubidge B. 1995. Biostratigraphy of the Beaufort Group (Karoo Sequence), South Africa. *Geol. Surv. S. Afr. Biostratigr.* 1:1–43.

Sepkoski J. 1984. A model of Phanerozoic taxonomic diversity. *Paleobiology* 10:246–267.

Smith R. 1990. A review of stratigraphy and sedimentary environments of the Karoo Basin of South Africa. *J. Afr. Earth Sci.* 10:117–137.

Smith R. 1995. Changing fluvial environments across the Permian-Triassic boundary in the Karoo Basin, South Africa and possible causes of tetrapod extinctions. *Palaeogeogr. Palaeoclimatol. Palaeoecol.* 117:81–104.

Stanley S. 1987. *Extinctions*. New York: W. H. Freeman.

Stanley S, Yang X. 1994. A double mass extinction at the end of the Paleozoic Era. *Science* 26:1340–1344.

Stuart C, Stuart T. *A Field Guide to the Larger Animals of Africa*. Cape Town: Struik Publishers.

Teichert C. 1990. The end-Permian extinction. Pp. 161–190 in *Global Events in Earth History*, E Kauffman, O Walliser, eds. New York: Springer-Verlag.

# DINOSAURS AND BIRDS (CHAPTERS 6, 7, AND 8)

Avery RA. 1982. Field studies of body temperatures and thermoregulation. Pp. 93–166 in *Biology of the Reptilia*, vol. 12, C Gans, HF Pough, eds. London: Academic Press.

Baker MA. 1982. Brain cooling in endotherms in heat and exercise. *Annu. Rev. Physiol.* 44:85–96.

Bakker RT. 1968. The superiority of dinosaurs. *Discovery* 3:11–12.

Bakker RT. 1971. Dinosaur physiology and the origin of mammals. *Evolution* 25:636–658.

Bakker RT. 1975. Dinosaur renaissance. *Sci. Am.* 232:58–78.

Bakker RT. 1980. Dinosaur heresy—dinosaur renaissance. Pp. 351–462 in *A Cold Look at the Warm-blooded Dinosaurs*, RDK Thomas, EC Olson, eds. Boulder, CO: Westview.

Bakker RT. 1986. *The Dinosaur Heresies*. New York: Morrow.

Bang B. 1971. Functional anatomy of the olfactory system in 23 orders of birds. *Acta Anat. (Basel)* 79:1–71.

Barrick RE, Showers WJ. 1994. Thermophysiology of *Tyrannosaurus rex*: evidence from oxygen isotopes. *Science* 265:222–224.

Barsbold R. 1983. Carnivorous dinosaurs from the Cretaceous of Mongolia: the joint Soviet-Mongolian palaeontological expedition. *Transactions* 19:1–117.

Bellairs ADA, Jenkin CR. 1960. The skeleton of birds. Pp. 241–300 in *Biology and Comparative Physiology of Birds*, AJ Marshall, ed. New York: Academic Press.

Bennett AF. 1973. Blood physiology and oxygen transport during activity in two lizards, *Varanus gouldii* and *Sauromalus hispidus*. *Comp. Biochem. Physiol.* 46A:673–690.

Bennett AF. 1973. Ventilation in two species of lizards during rest and activity. *Comp. Biochem. Physiol.* 46A:653–671.

Bennett AF. 1982. The energetics of reptilian activity. Pp. 155–199 in *Biology of the Reptilia*, vol. 13, C Gans, HF Pough, eds. New York: Academic Press.

Bennett AF. 1991. The evolution of activity capacity. *J. Exp. Biol.* 160:1–23.

Bennett AF, Dalzell B. 1973. Dinosaur physiology: a critique. *Evolution* 27:170–174.

Bennett AF, Dawson WR. 1976. Metabolism. Pp. 122–127 in *Biology of the Reptilia*, vol. 5, C Gans, WR Dawson, eds. New York: Academic Press.

Bennett AF, Ruben JA. 1979. Endothermy and activity in vertebrates. *Science* 206:649–654.

Bennett AF, Seymour RS, Webb GJW. 1985. Mass dependence of anaerobic metabolism and acid-base disturbance during activity in the saltwater crocodile, *Crocodylusporosus*. *J. Exp. Biol.* 118:161–171.

Bernstein MH, et al. 1984. Extrapulmonary gas exchange enhances brain oxygen in pigeons. *Science* 226:564–566.

Brackenbury JH. 1987. Ventilation of the lung-air sac system. Pp. 36–69 in *Bird Respiration*, TJ Sellers, ed. Boca Raton, FL: CRC Press.

Britt BB, Makovicky PJ, Gauthier J, Bonde N. 1998. Postcranial pneumatization in *Archaeopteryx*. *Nature* 395:374–376.

Brooke M, Birkhead TR. 1991. *The Cambridge Encyclopedia of Ornithology*. Cambridge, UK: Cambridge University Press.

Burnham DA, Derstler KL, Currie PJ, Bakker RT, Zhou Z-H, Ostrom JH. 2000. Remarkable new birdlike dinosaur (Theropoda: Maniraptora) from the Upper Cretaceous of Montana. *Univ. Kans. Paleontol. Contrib.* 13:1–14.

Carrier DR, Farmer CG. 2000. The evolution of pelvic aspiration in archosaurs. *Paleobiology* 26:271–293.

Carrier DR, Farmer CG. 2000. The integration of ventilation on locomotion in archosaurs. *Am. Zool.* 40:87–100.

Case TJ. 1978. On the evolution and adaptive significance of postnatal growth rates in terrestrial vertebrates. *Q. Rev. Biol.* 53:243–282.

Chatterjee S. 1997. The beginnings of avian flight. Pp. 311–335 in *Dinofest International: Proceedings of a Symposium Sponsored by Arizona State University*, DL Wolberg, et al., eds. Philadelphia, PA: Academy of Natural Sciences.

Chen P-J, Dong Z-M, Zhen S-N. 1998. An exceptionally well-preserved theropod dinosaur from the Yixian formation of China. *Nature* 391:147–152.

Chinsamy A. 1994. Dinosaur bone histology: implications and inferences. Pp. 317–323 in *Dinofest*, Paleontological Society Special Publication No. 7, GD Rosenberg, D Wolberg, eds. Knoxville: University of Tennessee Press.

Chinsamy A, Dodson P. 1995. Inside a dinosaur bone. *Am. Sci.* 83:174–180.

Chinsamy A, Hillenius WJ. In press. Physiology of nonavian dinosaurs. In *The Dinosauria*, DB Weishampel, et al., eds. Berkeley: University of California Press.

Chinsamy A, Chiappe LM, Dodson P. 1994. Growth rings in Mesozoic birds. *Nature* 368:196–197.

Clark JM, Norell MA, Chiappe LM. 1999. An oviraptorid skeleton from the Late Cretaceous of Ukhaa Tolgod, Mongolia, preserved in an avian-like brooding position over an oviraptorid nest. *Am. Mus. Novit.* 3265:1–35.

Colbert EH, Mook CC. 1951. The ancestral crocodilian *Protosuchus*. *Bull. Am. Mus. Nat. Hist.* 97:143–182.

Coombs WP, Jr. 1989. Modern analogs for dinosaur nesting and parental behavior. Pp. 21–53 in *Paleobiology of the Dinosaurs*, JO Farlow, ed. Boulder, CO: Books on Demand.

Coombs WP, Jr., Maryanska T. 1990. Ankylosauria. Pp. 456–483 in *The Dinosauria*, DB Weishampel, et al., eds. Berkeley: University of California Press.

Cott HB. 1961. Scientific results of an inquiry into the ecology and economic status of the Nile crocodile (*Crocodylus niloticus*) in Uganda and Northern Rhodesia. *Trans. Zool. Soc. London* 29:211–357.

Cott HB. 1971. Parental care in the Crocodilia, with special reference to *Crocodylus niloticus* in Crocodiles. Pp. 166–180 in *Proceedings of the First Meeting of Crocodile Specialists*. Switzerland: International Union for the Conservation of Nature and Natural Resources.

Crush PJ. 1984. A late Upper Triassic Sphenosuchid crocodilian from Wales. *Palaeontology* 27:131–157.

Currie PJ. 1997. Theropoda. Pp. 731–737 in *Encyclopedia of Dinosaurs*, PJ Currie, K Padian, eds. San Diego, CA: Academic Press.

Dal Sasso C, Signore M. 1998. Exceptional soft-tissue preservation in a theropod dinosaur from Italy. *Nature* 392:383–387.

Dal Sasso C, Signore M. 1998. In situ preservation in *Scipionyx*. P. 23 in *Third European Workshop of Vertebrate Paleontology—Maastricht*, JWM Jagt, et al., eds. Maastricht, The Netherlands: Naturhistorisches Museum.

Darveau C-A, Suarez RK, Andrews RD, Hochachka PW. 2002. Allometric cascade as a unifying principle of body mass effects on metabolism. *Nature* 417:166–170.

Dietz MW, Drent RH. 1997. Effect of growth rate and body mass on resting metabolic rate in galliform chicks. *Physiol. Zool.* 70:493–501.

Dodson P. 2000. Origin of birds: the final solution? *Am. Zool.* 40:504–512.

Dodson P, Currie PJ. 1990. Neoceratopsia. Pp. 593–618 in *The Dinosauria*, DB Weishampel, et al., eds. Berkeley: University of California Press.

Dong Z-M, Currie PJ. 1998. On the discovery of an oviraptorid skeleton on a nest of eggs at Bayan Mandahu, Inner Mongolia, People's Republic of China. *Can. J. Earth Sci.* 33:631–636.

Drent R. 1972. The natural history of incubation. Pp. 262–311 in *Breeding Biology of Birds*, DS Farner, ed. Washington, DC: National Academy of Sciences.

Duncker H-R. 1971. The lung air sac system of birds. *Adv. Anat. Embryol. Cell Biol.* 45:1–171.

Dunker H-R. 1972. Structure of avian lungs. *Respir. Physiol.* 14:44–63.

Dunker H-R. 1974. Structure of the avian respiratory tract. *Respir. Physiol.* 22:1–19.

Dunker H-R. 1978. General morphological principles of amniote lungs. Pp. 2–15 in *Respiratory Function in Birds, Adult and Embryonic*, J Piiper, ed. Berlin: Springer-Verlag.

Dunker H-R. 1989. The lung air sac system of birds. Pp. 39–67 in *Form and Function in Birds*, vol. 1, AS King, J McLelland, eds. New York: Academic Press.

Else PL, Hulbert AJ. 1981. Comparison of the "mammal machine" and the "reptile machine": energy production. *Am. J. Physiol.* 240:R3–R9.

Farlow JO. 1990. Dinosaur energetics and thermal biology. Pp. 43–55 in *The Dinosauria*, DB Weishampel, et al., eds. Berkeley: University of California Press.

Farlow JO, Dodson P, Chinsamy A. 1995. Dinosaur biology. *Annu. Rev. Ecol. Syst.* 26: 445–471.

Farmer CG, Carrier DR. 2000. Pelvic aspiration in the American alligator (*Alligator mississippiensis*). *J. Exp. Biol.* 203:1679–1687.

Farmer CG, Carrier DR. 2000. Ventilation and gas exchange during treadmill locomotion in the American alligator (*Alligator mississippiensis*). *J. Exp. Biol.* 203:1671–1678.

Fedde MR. 1987. Respiratory muscles. Pp. 3–37 in *Bird Respiration*, vol. 1, TJ Sellers, ed. Boca Raton, FL: CRC Press.

Feduccia A. 1999. Accommodating the clado-1,2,3p2,3,4 gram. *Proc. Natl. Acad. Sci. USA* 96:4740–4742.

Feduccia A. 1999. *The Origin and Evolution of Birds*, 2nd ed. New Haven, CT: Yale University Press.

Fisher PE, Russell DA, Stoskopf MK, Barrick RE, Hammer M, Kuzmitz AA. 2000. Cardiovascular evidence for an intermediate or higher metabolic state in an ornithsichian dinosaur. *Science* 288:503–505.

Forster CA. 1990. The postcranial skeleton of the ornithopod dinosaur *Tenontosaurus tilletti*. *J. Vertebr. Paleontol.* 10:273–294.

Frair W, Ackman RG, Mrosovsky N. 1972. Body temperature of *Dermochelys coriacea*: warm turtle from cold water. *Science* 177:791–793.

Franklin C, Seebacher F, Grigg GC, Axelsson M. 2000. At the crocodilian heart of the matter. *Science* 289:1687–1688.

Frey E. 1988. The carrying system of crocodilians: a biomechanical and phylogenic analysis. *Stuttg. Beitr. Naturkd.* A426:1–60.

Gans C. 1996. An overview of parental care among the Reptilia. *Adv. Study. Behav.* 25: 145–157.

Gans C, Clark B. 1976. Studies on the ventilation of *Caiman crocodilus* (Reptilia: Crocodilia). *Respir. Physiol.* 26:285–301.

Geist NR. 2000. Nasal respiratory turbinate function in birds. *Physiol. Biochem. Zool.* 73:581–589.

Geist NR, Feduccia A. 2000. Gravity-defying behaviors: identifying models for protoaves. *Am. Zool.* 40:664–675.

Geist NR, Jones TD. 1996. Juvenile skeletal structure and the reproductive habits of dinosaurs. *Science* 272:712–714.

Gill FB. 1994. *Ornithology*, 2nd ed. New York: W. H. Freeman.

Goodrich ES. 1930. *Studies on the Structure and Development of Vertebrates.* London: Macmillan.

Greenberg N. 1980. Physiological and behavioral thermoregulation in living reptiles. Pp. 141–166 in *A Cold Look at the Warm-blooded Dinosaurs*, RDK Thomas, EC Olson, eds. Boulder, CO: Westview.

Gregory WK. 1951. *Evolution Emerging.* New York: Macmillan.

Grenard S. 1991. *Handbook of Alligators and Crocodiles.* Malabar, FL: Krieger.

Hamilton WJ, Heppner F. 1966. Radiant solar energy and the function of black homeotherm pigmentation: a hypothesis. *Science* 155:196–197.

Harwell A, van Leer D, Ruben JA. 2002. New evidence for hepatic-piston breathing in theropods. *J. Vertebr. Paleontol.* 22:63A.

Hemmingsen AM. 1960. Energy metabolism as related to body size and respiratory surfaces and its evolution. *Rep. Steno. Mem. Hosp.* 9:7–110.

Hengst R. 1998. Ventilation and gas exchange in theropod dinosaurs. *Science* 281:47–48.

Hicks JW, Farmer CG. 1998. Ventilation and gas exchange in theropod dinosaurs. *Science* 281:47–48.

Hicks JW, Farmer CG. 1999. Gas exchange potential in reptilian lungs: implications for the dinosaur-avian connection. *Respir. Physiol.* 117:73–83.

Hildebrand M, Goslow GE, Jr. 2001. *Analysis of Vertebrate Structure.* New York: Wiley.

Hillenius WJ. 1992. The evolution of nasal turbinates and mammalian endothermy. *Paleobiology* 18:17–29.

Hillenius WJ. 1994. Turbinates in therapsids: evidence for Late Permian origins of mammalian endothermy. *Evolution* 48:207–229.

Hofstetter R, Gasc J-P. 1969. Vertebrae and ribs of modern reptiles. Pp. 201–310 in *Biology of the Reptilia*, vol. 1, Morphology A, C Gans, et al., eds. London: Academic Press.

Horner CC, Horner JR, Weishampel DB. 2001. Comparative internal cranial morphology of some hadrosaurian dinosaurs using computerized tomographic x-ray analysis and rapid prototyping. *J. Vertebr. Paleontol.* 21:64A.

Horner JR. 2000. Dinosaur reproduction and parenting. *Annu. Rev. Earth Planet. Sci.* 28:19–45.

Horner JR, de Ricqles AJ, Padian K. 2000. Long bone histology of the hadrosaurid dinosaur Maisaura peeblesorum: growth dynamics and physiology based on an ontogenetic series of skeletal elements. *J. Vertebr. Paleontol.* 20:115–129.

Horner JR, Padian K, de Ricqles AJ. 2001. Comparative osteohistology of some embryonic and perinatal archosaurs: developmental and behavioral implications for dinosaurs. *Paleobiology* 27:39–58.

Hutchinson JR. 2001. The evolution of pelvic osteology and soft tissues on the line to extant birds (Neorithes). *Zool. J. Linn. Soc.* 131:123–168.

Ingelstedt S. 1956. Studies on the conditioning of air in the respiratory tract. *Acta Otolaryngol.* 131:1–79.

Jackson DC, Schmidt-Neilsen K. 1964. Countercurrent heat exchange in the respiratory passages. *Proc. Natl. Acad. Sci. USA* 51:1192–1197.

Jones TD, Ruben JA. 2001. Respiratory structure and function in theropod dinosaurs and some related taxa. Pp. 443–461 in *Proceedings of the International Symposium in Honor of John H. Ostrom (February 13–14, 1999): New Perspectives on the Origins and Evolution of Birds*, J Gauthier, LF Gall, eds. New Haven, CT: Yale University Press.

Jones TD, Ruben JA, Martin LD, Kurochkin EN, Feduccia A, Maderson PFA, Hillenius WJ, Geist NR. 2000. Nonavian feathers in a Late Triassic Archosaur. *Science* 288:2202–2205.

Jones TD, Ruben JA, Maderson PFA, Martin LD. 2001. Longisquama fossil and feather morphology. *Science* 291:1899–1902.

Kemp TS. 1988. Haeomothermia or Archosauria? The interrelationships of mammals, birds, and crocodiles. *Biol. J. Linn. Soc.* 92:67–104.

King AS. 1966. Structural and functional aspects of the avian lung and air sacs. *Int. Rev. Gen. Exp. Zool.* 2:171–267.

King AS, King DZ. 1979. Avian morphology. Pp. 1–38 in *Form and Function in Birds*, AS King, J McLelland, eds. London: Academic Press.

Kolodny Y, Lutz B, Sander M, Clemens WA. 1996. Dinosaur bones: fossils or pseudomorphs? The pitfalls of physiology reconstruction from apatitic fossils. *Palaeogeogr. Palaeoclimatol. Palaeoecol.* 126:161–167.

Lang JW. 1989. Social behavior. Pp. 102–117 in *Crocodiles and Alligators*, CA Ross, ed. New York: Facts-on-File.

Lustick S, Talbot S, Fox E. 1970. Absorption of radiant energy in red-winged blackbirds (*Agelaius phoeniceus*). *Condor* 72:471–473.

Maina JN, Africa M. 2000. Inspiratory aerodynamic valving in the avian lung: functional morphology of the extrapulmonary primary bronchus. *J. Exp. Biol.* 203:2865–2876.

Marsh OC. 1883. Principal characters of American Jurassic dinosaurs. VI. Restoration of *Brontosaurus*. *Am. J. Sci.* 26:81–85.

Martill DM, Frey E, Sues H-D, Cruickshank ARI. 2000. Skeletal remains of a small theropod dinosaur with associated soft structures from the Lower Cretaceous Santana Formation of northeastern Brazil. *Can. J. Earth Sci.* 37:891–900.

Martin LD. 1991. Mesozoic birds and the origin of birds. Pp. 485–540 in *Origins of the Higher Groups of Tetrapods: Controversy and Consensus*, H-P Schultze, L Trueb, eds. Ithaca, NY: Comstock.

Martoff BS, Palmer WM, Bailey JR, Harrison JRI. 1980. *Amphibians and Reptiles of the Carolinas and Virginia*. Chapel Hill: University of North Carolina.

Maryanska T, Osmólska H, and Wolsan M. 2002. Avialan status for Oviraptorosauria. *Acta Palaeontol. Pol.* 47:97–116.

McIntosh JS. 1990. Sauropoda. Pp. 345–401 in *The Dinosauria*, DB Weishampel, et al., eds. Berkeley: University of California Press.

McLelland J. 1989. Anatomy of lungs and air sacs. Pp. 221–279 in *Form and Function in Birds*, vol. 4, AS King, J McLelland, eds. London: Academic Press.

McNab BK, Auffenberg WA. 1976. The effect of large body size on temperature regulation of the Komodo dragon, *Varanus komodoensis*. *Comp. Biochem. Physiol.* 55A:345–350.

Nagy KA. 1987. Field metabolic rates and food requirement scaling in mammals and birds. *Ecol. Monogr.* 57:111–128.

Norell M, Clark JM. 1996. Dinosaurs and their youth. *Science* 273:165–168.

Norell MA. 1995. Origins of the feathered nest. *Nat. Hist.* 104:58–61.

Norell MA, Makovicky PJ. 1997. Important features of the dromaeosaur skeleton: information from a new specimen. *Am. Mus. Novit.* 3215:1–28.

Norell MA, Makovicky PJ. 1999. Important features of the dromaeosaur skeleton. II. Information from newly collected specimens of *Velociraptor mongoliensis*. *Am. Mus. Novit.* 3282:1–44.

Norell MA, Clark JM, Chiappe LM, Dashzeveg D. 1995. A nesting dinosaur. *Nature* 378:774–776.

Norman DB. 1980. On the ornithischian dinosaur *Iguanodon bernissartensis* from the Lower Cretaceous of Bernissart (Belgium). *Inst. R. Sci. Nat. Belg. Mem.* 178:1–103.

Novas FE. 1993. New information on the systematics and postcranial skeleton of *Herrerasaurus ischigualastensis* (Theropoda: Herrerasauridae) from the Ischigualasto formation (Upper Triassic) of Argentina. *J. Vertebr. Paleontol.* 13:400–423.

O'Conner MP, Dodson P. 1999. Biophysical constraints on the thermal ecology of dinosaurs. *Paleobiology* 25:341–368.

Ohmart RD, Lasiewski RC. 1971. Roadrunners: energy conservation by hypothermia and absorption of sunlight. *Science* 172:67–69.

Olson JM. 1992. Growth, the development of endothermy, and the allocation of energy in red-winged blackbirds (*Agelaius pheoniceus*) during the nestling period. *Physiol. Zool.* 65:125–152.

Ostrom JH. 1969. A new theropod dinosaur from the Lower Cretaceous of Montana. *Postilla* 128:1–17.

Ostrom JH. 1991. The question of the origin of birds. Pp. 467–484 in *Origins of the Higher Groups of Tetrapods: Controversy and Consensus*, H-P Schultze, L Trueb, eds. Ithaca, NY: Comstock.

Padian K, Horner JR. 2002. Typology versus transformation in the origin of birds. *Trends Ecol. Evol.* 17:120–124.

Paladino FV, O'Conner MP, Spotila JR. 1990. Metabolism of leatherback turtles, gigantothermy, and thermoregulation of dinosaurs. *Nature* 344:858–860.

Paladino FV, Spotila JR, Dodson P. 1997. A blueprint for giants: modeling the physiology of large dinosaurs. Pp. 491–504 in *The Complete Dinosaur*, JO Farlow, MK Brett-Surnam, eds. Bloomington: Indiana University Press.

Paul GS. 2001. Were the respiratory complexes of predatory dinosaurs like crocodilians or birds? Pp. 463–482 in *Proceedings of the International Symposium in Honor of John H. Ostrom (February 13–14, 1999): New Perspectives on the Origins and Evolution of Birds*, J Gauthier, LF Gall, eds. New Haven, CT: Yale Peabody Museum of Natural History, Yale University.

Pearson OP. 1954. Habits of the lizard *Liolaemus multiformis multiformis* at high latitudes in southern Peru. *Copeia* 1954:111–116.

Perez-Moreno BP, Sanz JL, Buscalioni AD, Moratella JJ, Ortega F, Rasskin-Gutman D. 1994. A unique multi-toothed ornithomimisaur dinosaur from the Lower Cretaceous of Spain. *Nature* 370:363–367.

Perry SF. 1983. Reptilian lungs: functional anatomy and evolution. *Adv. Anat. Embryol. Cell Biol.* 79:1–83.

Perry SF. 1989. Mainstreams in the evolution of vertebrate respiratory structures. Pp. 1–67 in *Form and Function in Birds*, vol. 4, AS King, J McLelland, eds. London: Academic Press.

Perry SF. 1992. Gas exchange strategies in reptiles and the origin of the avian lung. Pp. 149–167 in *Physiological Adaptations in Vertebrates: Respiration, Circulation, and Metabolism*, SC Wood, et al., eds. New York: Marcel Dekker.

Perry SF. 2001. Functional morphology of the reptilian and avian respiratory systems and its implications for theropod dinosaurs. Pp. 429–441 in *Proceedings of the International Symposium in Honor of John H. Ostrom (February 13–14, 1999): New Perspectives on the Origins and Evolution of Birds*, J Gauthier, LF Gall, eds. New Haven, CT: Yale Peabody Museum of Natural History, Yale University.

Perry SF, Duncker H-R. 1980. Interrelationships of static mechanical factors and anatomical structure in lung evolution. *J. Comp. Physiol. B* 138:321–334.

Perry SF, Reuter C. 1999. Hypothetical lung structure of *Brachiosaurus* (Dinosauria: Sauropoda) based on functional constraints. *Mitt. Mus. Nat. Berl. Geowissenschaftliche Reihe* 2:75–79.

Peters DS, Gorgner E. 1992. A comparative study on the claws of *Archaeopteryx*. Pp. 29–37 in *Proceedings of the Second International Symposium of Avian*, K Campbell, ed. Los Angeles, CA: Los Angeles Museum of Natural History Press.

Pough FH, Andrews RM, Cadle JE, Crump ML, Savitzky AH, Wells KD. 1998. *Herpetology*. Upper Saddle River, NJ: Prentice Hall.

Proctor DF, Andersen I, Lundqvist GR. 1977. Human nasal mucosal function at controlled temperatures. *Respir. Physiol.* 30:109–124.

Prum RO. 2001. *Longisquama* fossil and feather morphology. *Science* 291:1899–1900.

Randall D, Burggren WW, French K. 2002. *Eckert's Animal Physiology: Mechanisms and Adaptations*, 5th ed. New York: W. H. Freeman.

Randolph SE. 1994. The relative timing of the origin of flight and endothermy: evidence from the comparative biology of birds and mammals. *Zool. J. Linn. Soc.* 112: 389–397.

Reid REH. 1997. Dinosaurian physiology: the case for "intermediate" physiology. Pp. 449–473 in *The Complete Dinosaur*, JO Farlow, MK Brett-Surnam, eds. Bloomington: Indiana University Press.

Ricklefs RE. 1979. Adaptation, constraint, and compromise in avian post-natal development. *Biol. Rev. Camb. Phil. Soc.* 54:269–290.

Romer AS. 1956. *Osteology of the Reptiles*. Chicago, IL: University of Chicago Press.

Rowe T, McBride EF, Sereno PC. 2001. Dinosaur with a heart of stone. *Science* 291:783.

Royer DL, Osborne CP, Beerling DJ. 2002. High $CO_2$ increases the freezing sensitivity of plants: implications for paleoclimatic reconstructions from fossil floras. *Geology* 30:963–966.

Ruben JA. 1991. Reptilian physiology and the flight capacity of *Archaeopteryx. Evolution* 45:1–17.

Ruben JA. 1995. The evolution of endothermy in mammals and birds: from physiology to fossils. *Annu. Rev. Physiol.* 57:69–95.

Ruben JA. 1996. Evolution of endothermy in mammals, birds and their ancestors. Pp. 347–376 in *Animals and Temperature: Phenotypic and Evolutionary Adaptation,* IA Johnston, AF Bennett, eds. Cambridge, UK: Cambridge University Press.

Ruben JA, Jones TD. 2000. Selective factors for the origin of fur and feathers. *Am. Zool.* 40:585–596.

Ruben JA, Hillenius WJ, Geist NR, Leitch A, Jones TD, Currie PJ, Horner JR, Espe G III. 1996. The metabolic status of some Late Cretaceous dinosaurs. *Science* 273:1204– 1207.

Ruben JA, Jones TD, Geist NR, Hillenius WJ. 1997. Lung structure and ventilation in theropod dinosaurs and early birds. *Science* 278:1267–1270.

Ruben JA, Leitch A, Hillenius WJ, Geist NR, Jones TD. 1997. New insights into the metabolic physiology of dinosaurs. Pp. 505–518 in *The Complete Dinosaur,* JO Farlow, MK Brett-Surnam, eds. Bloomington: Indiana University Press.

Ruben JA, Jones TD, Geist NR, Hillenius WJ. 1998. Ventilation and gas exchange in theropod dinosaurs. *Science* 281:47–48.

Ruben JA, Dal Sasso C, Geist NR, Hillenius WJ, Jones TD, Signore M. 1999. Pulmonary function and metabolic physiology of theropod dinosaurs. *Science* 283:514–516.

Ruxton GD. 2000. Statistical power analysis: application to an investigation of dinosaur thermal physiology. *J. Zool.* 252:239–241.

Scheid P, Piiper J. 1989. Respiratory mechanics and air flow in birds. Pp. 369–388 in *Form and Function in Birds,* vol. 4, AS King, J McLelland, eds. New York: Academic Press.

Schmidt-Neilsen K. 1971. How birds breathe. *Sci. Am.* 225:72–79.

Schmidt-Neilsen K. 1984. *Scaling.* Cambridge, UK: Cambridge University Press.

Schmidt-Neilsen K. 1990. *Animal Physiology: Adaptation and Environment,* 4th ed. Cambridge, UK: Cambridge University Press.

Seebacher F, Grigg GC, Beard LA. 1999. Crocodiles as dinosaurs: behavioral thermo- regulation in very large ectotherms leads to high and stable body temperatures. *J. Exp. Biol.* 202:77–86.

Sereno PC, Chao S. 1988. *Psittacosaurus meileyingensis. J. Vertebr. Paleontol.* 8:353–365.

Seymour RS. 1979. Dinosaur eggs: gas conductance through the shell, water loss during incubation and clutch size. *Paleobiology* 5:1–11.

Shine R. 1988. Parental care in reptiles. Pp. 275–329 in *Biology of the Reptilia,* C Gans, RB Huey, eds. New York: Liss.

Spotila JR, O'Conner MP, Dodson P, Paladino FV. 1991. Hot and cold running dinosaurs: body size, metabolism and migration. *Mod. Geol.* 16:203–227.

Standora EA, Spotila JR, Keinath JA, Shoop CR. 1984. Body temperatures, diving cycles, and movement in the subadult leatherback turtle, *Dermochelys coriacea. Herpetologica* 40:169–176.

Starck JM. 1996. Comparative morphology and cytokinetics of skeletal growth in hatchlings of altricial and precocial birds. *Zool. Anz.* 235:53–75.

Starck JM, Chinsamy A. 2002. Bone microstructure and developmental plasticity in birds and other dinosaurs. *J. Morphol.* 254:232–246.

Thomas DW, Bosque C, Arends A. 1993. Development of thermoregulation and energetics of nestling oilbirds (*Steatornis caripensis*). *Physiol. Zool.* 66:322–348.

Upchurch P. 1995. The evolutionary history of sauropod dinosaurs. *Philos. Trans. R. Soc. London B* 349:365–390.

Varricchio DJ, Jackson F, Borkowski JJ, Horner JR. 1997. Nest and egg clutches of the dinosaur *Troodon formosus* and the evolution of avian reproductive traits. *Nature* 385:247–250.

Visser GH, Ricklefs RE. 1993. Development of temperature regulation in shorebirds. *Physiol. Zool.* 66:771–792.

von Koenigswald W, Storch G, Richter G. 1988. Ursprüngliche "Insectenfresser," extravagante Igel und Langfinger. Pp. 159–177 in *Messel: ein Schaufenster in die Geschichte der Erde und des Leben*, S Schaal, W Zeigler, eds. Frankfurt: Waldemar Kramer.

Weishampel DB. 1981. Acoustic analysis of potential vocalization in lambeosaurine dinosaurs (Reptilia: Ornithischia). *Paleobiology* 7:252–261.

Weishampel DB. 1981. The nasal cavity of lambeosaurine hadrosaurids (Reptilia: Ornithischia). *J. Paleontol.* 55:1046–1057.

Weishampel DB. 1997. Dinosaurian cacophony. *BioSci* 47:150–159.

Weishampel DB, Norman DB, Grigorescu D. 1993. *Telmatosaurus transsylvaticus* from the Late Cretaceous of Romania: the most basal hadrosaurid dinosaur. *Paleontology* 36:361–385.

Welty JC, Baptista L. 1988. *The Life of Birds*. Fort Worth, TX: Saunders College.

Westoby M, Leishman MR, Lord JM. 1995. On misinterpreting the "phylogenetic correction." *J. Ecol.* 83:531–534.

Wettstein O. 1931. Rhynchocephalia. Pp. 1–235 in *Handbuch der Zoologie*, W Kukenthal, T Krumbach, eds. Berlin: De Gruyter.

Wettstein O. 1976. Circulation. Pp. 275–334 in *Biology of the Reptilia*, vol. 5, C Gans, WR Dawson, eds. New York: Academic Press.

Witmer LM. 1995. Homology of facial structures in extant archosaurs (birds and crocodilians), with special reference to paranasal pneumaticity and nasal conchae. *J. Morphol.* 225:269–327.

Witmer LM. 1997. The evolution of the antorbital cavity of archosaurs: a study in soft-tissue reconstruction in the fossil record with an analysis of the function of pneumaticity. *J. Vertebr. Paleontol.* 17:1–73.

Witmer LM. 1999. Nasal conchae and blood supply in some dinosaurs: physiological implications. *J. Vertebr. Paleontol.* 19:85A.

Xu X, Tang Z-L, Wang X-L. 1999. A therizinosauroid dinosaur with integumentary structures from China. *Nature* 399:350–354.

Young C-C. 1947. On *Lufengosaurus magnus* (sp. nov.) and additional finds of *Lufengosaurus heunei* young. *Palaeontol. Sin.* 12:1–53.

APPENDIX

RESPIRATORY SYSTEMS AMONG
VARIOUS ANIMAL GROUPS

## Respiratory Systems Among Various Animal Groups

| Taxon | Medium | Type | Absorption surface |
|---|---|---|---|
| Humans | Air | Pump lung | Complex alveoli |
| Birds | Air | Pump lung with added air sac | Simple alveoli in lung, and internal surface of air sacs |
| Lizards | Air | Pump lung | Simple alveoli |
| Crocodiles | Air | Pump lung | Simple alveoli |
| Saurischian dinosaurs | Air | Pump lung with added air sac | ? |
| Fish | Water | Counter current gills | Gill surface |
| Bivalve mollusks | Water | Medium pressure | Pump gills |
| Chambered cephalopods | Water | Complex gills, high pressure pump gills | Gill surface |
| Brachiopods | Water | Low pressure pump gills | Lophophore surface, punctae (terebratulids) |
| Scleractinian corals | Water | No gill epithelial adsorption | Epithelium, internal mesentaries |
| Echinoids | Water | External gill, adsorption, tube feet | Gill surface |
| Bryozoans | Water | Lophophore | Epithelial absorption |
| Sponges | Water | No gills, medium pressure pump | Choanocyte cell surface |
| Tunicates | Water | Internal gill, medium pressure pump | Internal gill surface |

| Surface area/ volume of resp. organ | Blood flow across resp. organ | Source of pump | Chambers in heart | Blood pigment |
|---|---|---|---|---|
| High | High | Diaphragm breathing | Four | Hemoglobin |
| High | High | Ribcage, pelvic | Four | Hemoglobin |
| Low | Low | Diaphragm | Three | Hemoglobin |
| Low | Low | Pelvic pump | Three | Hemoglobin |
| ? | ? | Ribcage, pelvic | Four | ? |
|  | Low | Rib ventilation | Two |  |
|  | Low | Cilia in siphon | One | None |
|  | High | Adductor muscle contraction | One | Hemocyanin |
|  | Low | Cilia | None | None |
|  | None | None | None | None |
|  | Low | Water vascular system |  |  |
|  | Low | Cilia | None | None |
|  | None | Cilia | Flagellar beating | None |
|  | Low | Cilia |  |  |

# INDEX